Golf Course Tree Management

by
Sharon Lilly

Ann Arbor Press

Library of Congress Cataloging-in-Publication Data

Lilly, Sharon
    Golf course tree management / by Sharon Lilly ; edited by Jim
Skiera.
        p.   cm.
    Includes index.
    ISBN 1-57504-117-0
    1. Golf courses—Management. 2. Trees, Care of. I. Skiera, Jim.
II. Title.
    GV975.5.L55   1999
    796.352′06′8—dc21   98-49650
                                                                              CIP

ISBN 1-57504-117-0

ANN ARBOR PRESS
121 South Main Street, Chelsea, Michigan 48118
Ann Arbor Press is an imprint of Sleeping Bear Press, Inc.

PRINTED IN THE UNITED STATES OF AMERICA
10 9 8 7 6 5 4 3 2 1

Acknowledgments

This book would not have been possible without the technical and moral support of many individuals. Writing a book can be compared to a long journey, and the actual writing is just one step. Anyone who has ever been a part of the process can appreciate the frustrations and setbacks involved. I thank my family and friends for supporting me and for helping to bring this project to fruition.

The inspiration for this book came from several coworkers and business associates who help bridge the gap between arborists and golf course superintendents. Two key people who have been involved since the book's inception are Jim Skiera (The International Society of Arboriculture, Champaign, IL) and Chad Ritterbusch (Selz/Seabolt Communications, Chicago, IL). In addition, some of the early ideas for content came from Colin Bashford (CBA, Ltd., South Hampton, England).

I would also like to thank all of my colleagues who provided feedback through technical review of the text: Dr. Bonnie Appleton, (Hampton Roads Research Center, Virginia Beach, VA), Bill Baker (Recreation Centers of Sun City West, Riverside, CA), Dr. John Ball (South Dakota State University, Brookings, SD), Ed Brennan, (HortScience, Inc., Pleasanton, CA), Dr. William Chaney (Purdue University, West Lafayette, IN), Dr. Kim Coder (University of Georgia, Athens, GA), Tim Gamma (Gamma's Tree Service, St. Louis, MO), Bob Graunke (Eagle Ridge Inn & Resort, Galena, IL), Larry Hall (The Care of Trees, Wheeling, IL), Jim Ingram (Bartlett Tree Expert Co., Osterville, MA), John Lloyd (Ohio State University, Wooster, OH), Ken Meyer (Mayne Tree Expert Co., San Mateo, CA), Randy Miller (Pacific Corporation, Bountiful, UT), Gary Mullane (Low Country Tree Care, Hilton Head, SC), Ward Peterson (Davey Tree Expert Co., Kent, OH), Peter Sortwell (Arbor Care, San Jose, CA), Ted Stamen (Ted Stamen & Associates, Riverside, CA), and Dr. Gary Watson (The Morton Arboretum, Lisle, IL).

I gratefully acknowledge the contributions of the following people who provided photographs for use in this book: Dr. Jim Clark and Nelda Matheny (HortScience, Inc., Pleasanton, CA), Peter Sortwell, (Arbor Care, San Jose, CA), Tim Gamma (Gamma's Tree Service, St. Louis, MO), Dr. Alex Shigo (Shigo and Trees, Durham, NH), Bill Kruidenier and Jim Skiera (The International Society of Arboriculture, Champaign,

IL), Ian Bruce (Humber College, Ontario, Canada), The National Arborist Association (Amherst, NH), Ken Meyer (Mayne Tree Expert Co., San Mateo, CA), Dr. John Ball (South Dakota State University, Brookings, SD), Jim Chatfield (Ohio State University, Wooster, OH), Dr. Dave Shetlar (Ohio State University, Columbus, OH), The Ohio Nursery and Landscape Association (Columbus, OH), Dr. Donald Ham (Clemson University, Clemson, SC), Ken Palmer and Rip Tompkins (ArborMaster Training, Inc., Natick, MA), and Ann Sherrill (Sherrill Equipment Co., Greensboro, NC). Finally, I thank The International Society of Arboriculture and Mike Thomas (Thomas Graphics, Champaign, IL) for providing the illustrations for this book. If a picture is indeed worth a thousand words, my readers will be grateful for the more than one hundred illustrations and photographs in this book!

Foreword

Few things excite a golf course architect more than a challenging site. Rolling terrain...flowing streams...and, yes, standing trees all beckon to us as Mother Nature's own siren song.

While the game took root in Scotland, where windswept heather and gorse lend charm to the ancient links, Mother Nature has afforded a different set of assets to most of the world. Whether golf is played in Asia or America, it is typically trees which distinguish and characterize the playing ground.

We employ trees to provide strategy and beauty, to augment the golfer's enjoyment of the game. We use them to define fairways and create visual effect. At the same time, we use them for practical purposes like shading us from the blazing sunshine and screening property lines. Indeed, trees illustrate the highest responsibility of golf course architecture: to balance form and function.

But for all their beauty and value, trees can be problematic. They can make life miserable for the high-handicapper. They can suffocate the delicate life of a golf green. They compete fiercely with the turfgrass that superintendents work hard to groom.

Yet we invest—or at least we should!—in the care of trees precisely because we recognize their value. Club members and other players take pride in their oaks and pines. They fall in love with their trees, which is why they scream when cutting a tree down is proposed, even when it's for the health and well-being of the golf course and the other trees.

Golf Course Tree Management is intended to serve as a technical resource for those who manage golf courses and shoulder the responsibility of caring for trees. Its arrival is long overdue!

Within these pages is a wealth of practical information. Refer to it to understand the maintenance requirements for trees. Let it help you write the specifications for a tree maintenance plan at your course. Use it to make the most important decisions involving your trees.

Keep this book on your desk and let it serve you for years to come.

Bob Lohmann, ASGCA
President
American Society of Golf Course Architects

Table of Contents

1

The Value and Importance of Trees on Golf Courses

Golf courses are valuable open spaces in predominantly urban areas. Even nongolfers can appreciate a golf course for its rolling, green carpets of grass and manicured landscape. Golf courses are designed to be aesthetically pleasing, but their primary function is to serve as the arena for a sport. The turf is the playing surface of the game of golf; its role is clearly understood and undisputed. The other major elements of most golf courses in the United States are the trees. Many golfers don't realize the functional value of the trees on a golf course, although most appreciate their presence. Golf course superintendents may curse the trees for the problems trees cause in caring for the turf. But this is a love/hate relationship for the superintendents because they realize that trees are critical to the course.

The Play of the Game

The architectural design of most golf courses depends heavily on trees. Trees are used to direct the fairways, defining and controlling the line of play. They delineate boundaries from one fairway to the next, or areas that are out of play. Sometimes trees create a chute that golfers must negotiate to land in play. They create doglegs, increasing the challenge of the course. The height of their lowest branches may force a low shot, or the height of their canopy tops may allow golfers to clear them. Trees may be used to "guard" the greens, challenging even the most highly skilled. Trees force golfers to choose routes of play.

Trees provide targets for golfers to aim their shots. They assist in judging distances for deciding which club to use. They help players formulate their strategies and plan their shots. They improve the visibility of the ball in flight and provide references to find balls which have gone out of sight. Trees are sometimes used as natural yardage markers, adding to the character of the course.

Golf is a sport played predominantly in the open sunlight (at least by those competent in the game). Trees provide a welcome respite from the sun, when planted where golfers can take refuge between greens and tees. A course with no mature trees can be brutal on a hot, summer day.

Visual Enhancement

Trees visually enhance a golf course. They are the vital third dimension, defining spaces and giving a feeling of privacy and comfort. Trees

1.1. Trees are often used to direct the play of fairways.

1.2. Trees can be used to define doglegs.

1.3. Trees can be used to "guard" a green,
controlling the approach.

1.4. The placement of trees on fairways can be a large part of determining the skill level of the course.

1.5. Limbs that extend over fairways can be a play hazard, whether intentional or not. Superintendents should work with arborists in assessing pruning requirements.

create the outdoor "walls" of the landscape, around which the other elements flow. They are used to create backdrop effects around greens and tees. They provide depth perception for players aiming for greens. They tie the elements of the course together and provide a sense of proportion.

Trees are used to screen unwanted views of maintenance buildings, adjoining properties, and roadways. They screen passersby's views of the golfers, reducing motorists' temptation to honk their horns just as a player takes a backswing. They can also frame a view, such as the scenic view of a lake or the landscaped approach to the clubhouse.

Trees have an inherent aesthetic value. The majestic beauty of a mature, spreading oak tree is unrivaled in nature. A large, stately tree

1.6. Trees that overhang fairways often influence play strategy. Trees overhanging tees can lead to uneven turf wear.

demands respect for its age and imposing presence. Yet the delicate beauty of a small, cutleaf Japanese maple can make it the focal point of a landscape.

Trees are unique in that their character changes dramatically from season to season. Some produce a flower show that lights up a space and fills the air with the familiar scent of spring. The summer brings a dense canopy of foliage which shades the ground and cools the air. In temperate climates, the autumn change of color can draw visitors from miles away. Then winter reveals the trees' structural scaffolds and the texture of the bark. The turf may fade from green to golden and may be blanketed with snow, but the trees are ever changing in color, character, and dimension.

1.7. Although shade can create problems for turf management, it can also be a blessing to golfers awaiting their turn.

1.8. Trees are often used to frame greens. This provides a
visual backdrop and aids players in following
the flight of golf balls.

1.9. Trees separate fairways, mitigating liability and
providing enclosure.

Other Benefits

The benefits of trees to a golf course are not limited to their aesthetic contributions or function in the play of the game. One area where trees play an important role is in liability mitigation. Trees provide barriers that can prevent errant golf shots from golfers on other fairways. They help protect cars on adjacent roadways and houses on adjacent properties.

Trees increase the property values, not only on the course, but also on the adjacent properties. The property value of a lot with mature trees can be worth up to 20 percent more than one without trees. If those trees frame a view of a well-kept golf course, the lot will be in high demand.

Trees add environmental value to the golf course as well. Trees can be strategically placed to block the wind. A deciduous tree can shade an area in the summer, while allowing the warming rays of the sun to pass through in the winter. A thick planting of evergreens can help with noise abatement. Trees contribute to air quality by taking in carbon dioxide and giving off oxygen. They also collect particle pollutants from the air.

1.10. Often a large, majestic tree will become the course's "signature" tree, sometimes even providing the course's name.

1.11. Trees can be used to divide fairways,
allowing efficient use of the land.

1.12. This tree in front of the clubhouse can be the most-valued tree
on the course. Proper care and maintenance can help
ensure its survival for many decades.

Trees may serve as a habitat for a variety of birds and small animals. Conservation issues are becoming more important as urban areas expand. Golf courses can be wildlife sanctuaries, and can play an important role in the conservation of natural resources. Many progressive courses are now participating in the Audubon Cooperative Sanctuary Program, which is an effort to work with golf course managers to implement ecologically sound conservation programs. Trees are an important part of this effort.

Financial Value

There are ways of appraising the value of an individual tree or group of trees. The Council of Tree and Landscape Appraisers has developed a formula that consulting arborists use to estimate the value of a tree based on its size, species, location, and condition. A large, healthy tree that is key to its location and serves a function may easily be appraised at more than $10,000. If the tree in question is the signature tree of the course, or if it "guards" the corner of a dogleg, its value to the course

1.13. Limbs that extend over greens can affect the play as well as the health of the turf.

may be even higher. The loss of such a tree could be devastating and could require redesigning an entire fairway.

Other techniques of valuing trees are based on the replacement value or the cost of cure (to restore the landscape if a tree or trees are lost). Golf courses sometimes hire a consulting arborist to appraise important trees on the course. This is a good idea, since it is much easier to document the appraisal if it is done while the tree is standing. Arborists are often called to assess the value of a tree that has been lost due to a storm or perhaps wrongfully cut down. Such appraisals are difficult to do and are more open to challenge once the tree is gone.

An appraiser can also estimate the value of all the trees on the golf course, if desired. This value can be a staggering figure, even if the trees are not appraised individually, but imagine the loss if a tornado or hurricane swept through and destroyed many of the trees on a golf course. Even a well-insured country club would have difficulty demonstrating the extent of their loss. And even if relatively large trees are planted in the rebuilding process, it will take generations before the beauty of the mature course is restored.

Golf courses typically spend thousands of dollars each year on new plantings alone. If you add to that the investment in maintenance and the appreciation with maturity, it is easy to see that the trees on a golf course represent a large capital asset. Although management tends to balk at the idea of devoting a significant portion of the maintenance budget to the trees, it must be viewed as an investment in that capital asset. When the maintenance figure is averaged out over the number of trees, then amortized over the life span of the trees, the expense seems minimal.

Maintaining and Increasing the Trees' Value

Most golf courses recognize that planting trees is a more or less continuous project. In some cases the tree planting has been extreme, driven by well-intentioned greens committees or memorial tree planting programs. Nevertheless, it is wise to maintain a wide variety of tree ages and maturities on the course. Trees will age and decline or become structurally hazardous, and it is a mistake to allow the course to mature without nurturing new generations.

Newly planted trees require a higher degree of care than established trees. They must be watered frequently and mulched to keep them alive and thriving in the first season. Later they will require formative pruning and perhaps fertilization. They must be protected against physical injury. They must be monitored for early signs of pest or disease problems, root health problems, or other indications of decline. Golf course superintendents understand these early needs and are accustomed to providing them.

There is a tendency, however, to take the established trees for granted, feeling that they don't require routine maintenance. What many golf course superintendents fail to realize is that the trees on a golf course are often surviving in a stressed environment. Most superintendents know that the trees can stress and inhibit the growth of turf, but this is actually a two-way street. Chapter 3 explains that trees and turf can be incompatible species, each inhibiting the other. When the problems that arise from trees and turf growing together in the same landscape are better understood, measures can be taken to reduce the stress and optimize growing conditions for each.

There are two aspects to the longevity of a tree on a golf course: health and structural security. Obviously, a healthier tree will have a longer service life on the course. Yet a tree can be relatively healthy and be totally unsafe. The same trees that reduce liability by preventing stray golf balls from hitting other golfers, passing motorists, or adjacent properties can be a liability themselves. Trees are assets that need to be maintained. Their health and structural stability can be maximized with due care.

Golf course superintendents as a group are perhaps the most highly educated professionals in the green industry, and justifiably, most of them have a very strong background in turf care. Few superintendents, however, have much knowledge of arboriculture. Since trees are a major element of a golf course, it is imperative that golf course superintendents learn more about tree care. This book is designed as a handbook to help golf course superintendents understand trees and their maintenance requirements. Recognizing that many golf course maintenance crews do not have the equipment, personnel, or training to do all of the tree care in-house, this text will aid in selecting a qualified arborist and in writing tree maintenance specifications.

Summary

1. Golf courses are designed to be aesthetically pleasing, but their primary function is to serve as a sport's arena. Although most golfers appreciate the presence of trees on golf courses, many

don't realize the function that the trees serve. Golf course superintendents may curse the trees for the problems they create in caring for the turf, but at the same time, they recognize how critical the trees are to the course.

2. The architectural design of most golf courses depends heavily on trees. Trees are used to direct fairways, delineate boundaries, create doglegs, and "guard" the greens. Golfers use the trees on the course to aim their shots.

3. Trees provide a respite from the hot sun. They provide a visual backdrop, defining spaces, and aiding golfers in seeing the ball in flight. Trees have an inherent aesthetic value, creating three-dimensional perspective and enhancing the landscape.

4. Trees can be used to reduce liability by creating a barrier to stop errant shots. They can protect adjacent property and roads, as well as adjoining fairways.

5. The trees on a golf course contribute environmentally by blocking wind, reducing noise, contributing to air quality, and providing a habitat for wildlife.

6. Trees add value to a golf course. As a group, they increase property values, yet each individual tree is a capital asset to be valued and managed.

2

Understanding How Trees Grow

It is unfathomable to imagine allowing a doctor to treat a patient if that doctor has not been thoroughly schooled in human anatomy and physiology, yet for centuries we have permitted the arboriculturally uneducated to perform tree surgery and administer treatments. The fact is, a large percentage of green industry professionals who care for trees are not arborists and may not understand even the most basic tenets of tree anatomy and physiology.

A skilled tree care provider learns how a tree grows in order to manage it in a way that supports its growth and development. Like physicians, arborists use knowledge of tree anatomy and physiology to diagnose problems, assess potential, and prescribe treatments, and, just as in medicine, prescription before diagnosis is malpractice. It is essential to evaluate tree biology before embarking on a program of care.

The study of tree biology is the study of the tree's structure and function, and the relationship between them. Anatomy and morphology are the studies of the component parts of the tree. Physiology is the study of the biological and chemical processes within these structures.

While researchers still have not unlocked all the secrets of the biology of a tree, they have gathered enough information about tree anatomy and physiology to fill volumes. Clearly, to care for trees on golf courses, one need not understand the biochemical intricacies of photosynthesis. It is, however, important to have a fundamental knowledge of what photosynthesis is, where it takes place, and how interruption of the process can affect the health of the tree. This short discussion is intended to be an elementary introduction to tree biology.

From the Ground Up

Trees are large, woody, long-lived perennials that fall into two classes, the angiosperms, with covered seeds (within an ovary), and the gymnosperms, with "naked" seeds. Angiosperms are the flowering plants, which include most deciduous trees and broad-leaved evergreens. Conifers, or cone-bearing plants, are in the gymnosperm class. Angiosperms are divided into two subclasses, the dicotyledons and the monocotyledons. The dicots include most tree species. Palms, however, are monocots and have a significantly different biology. The focus of this overview of tree biology is the structure and function of the dicotyledonous trees.

When the trunk of a tree has been cut it can be viewed in cross section. This provides the opportunity to learn about the anatomy of the stem tissues. Most people are familiar with the growth rings and know that counting them can give an approximate age of the tree. Fewer people understand the tissues and cell types that comprise the rings, how they are formed, or how they function.

Starting from the outside, the first thing encountered is the bark. The bark is actually a complex system of living and dead cells. The outer-

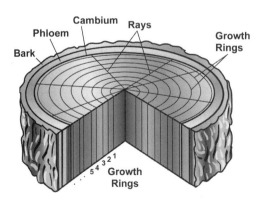

2.1. Cross section of wood showing growth rings and rays.

most cells are mostly thickened, dead corky cells. They protect the tree from water loss and mechanical injury such as might be caused by errant golf balls. Small openings, called lenticels, allow for exchange of gases, mostly oxygen and carbon dioxide. This tissue is produced by the cork cambium. Many types of bark develop with different species of trees. Beech trees have very smooth bark with little corky material. Cork oak produces thick layers of cork which is made into stoppers for wine and many other products.

The phloem tissues, sometimes referred to as the inner bark, lie just inside the outer bark. The phloem functions to translocate sugar and other materials throughout the tree. The phloem carries sugar products of photosynthesis from the leaves to the roots, but it also distributes these products for storage and consumption in other parts of the tree.

The vascular cambium is located just inside the phloem. It is not easy to see because it is only a few layers of cells in thickness. The cambium is meristematic tissue, which means it is a zone of rapid cell division.

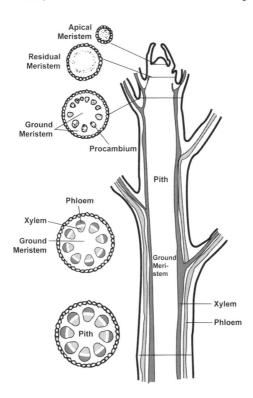

2.2. Longitudinal and cross sections through a shoot tip.

The cambium produces new cells that differentiate into specialized tissues. The cells produced toward the outside become phloem. Cells produced toward the inside become xylem. More xylem than phloem is generally produced, and old phloem is sloughed off each year. Thus the cambium moves outward as the tree grows.

The xylem is the wood of the tree, composed of both living and dead cells. It has four functions: conduction of water and mineral elements, support of the weight of the tree, storage of reserves, and defense against the spread of disease and decay. Not all of the conducting elements in the xylem transport water. In conifers, 2–12 outer rings may conduct water, while in some deciduous trees only the outermost ring conducts. The outer xylem rings that conduct water and contain some living cells comprise the sapwood. Increase in girth (diameter) of a tree is brought about primarily by the creation of a new layer of xylem each year. The familiar growth rings are visible because the relative size and density of the vascular tissues change as the growing season progresses. Cells produced in the early spring (earlywood) are larger in diameter than those produced toward the end of the growing season. Since some tropical

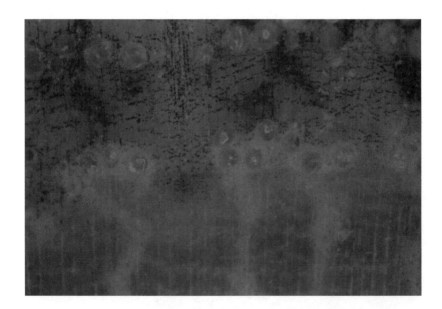

2.3. This photograph shows the large springwood vessels of the xylem. Starch (stained for visibility) is stored in nonwoody cells in the sapwood.

trees may grow more or less continuously, they may not produce visible growth rings.

Interspersed among the conducting elements of the phloem and xylem are thin-walled living cells known as parenchyma. These cells function as storage sites and may be active in moving materials in and out of the vascular elements of the phloem and xylem. Many parenchyma cells are aligned radially in layers that form the rays. Rays cross through many layers of phloem and xylem. They are the primary storage areas for starch. Rays also transport materials radially within the tree trunk and branches, and assist in restricting decay in wood tissue.

Although it varies with species and maturity, only about 10 percent of the cells in the sapwood are living. The vascular elements are composed of nonliving cells with thickened cell walls. As the tree matures and the xylem ages, the innermost cells die and become heartwood, which is physiologically inactive. Many compounds such as tannins, phenols, and terpenes are deposited in the heartwood and account for the darker color and decay resistance properties in some species. If the heartwood decays, the tree can still live with a hollow trunk because the living cells and the actively conducting vascular elements are in the sapwood.

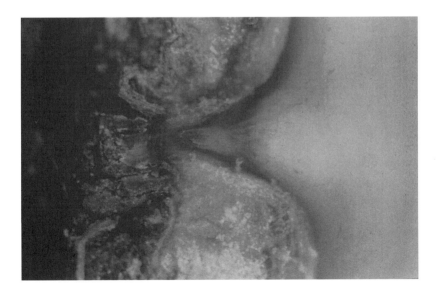

2.4. Trees are covered with millions of potential growth points that may be triggered to grow, especially if the tree is stressed.

As previously mentioned, trees increase in diameter through the meristematic activity of the cambium. Longitudinal growth of the shoots and roots occurs in meristem tissues located in the shoot tips and root tips. The bud located at the tip of a shoot is called the terminal, or apical bud. Buds that occur along the stem are called lateral, or axillary buds. Normally, the terminal bud is the most active on each branch or twig, while the axillary buds may be dormant. The growth of these axillary buds may be inhibited by the apical control of the terminal bud; that is, the terminal bud biochemically inhibits the growth and development of lateral buds on the same shoot. Removing the terminal bud in pruning can release dormant buds near the cut, leading to new shoot development.

Nodes are the areas on a twig where leaves and buds arise. The leaves and bud scales leave scars along the twig after the leaves have fallen. It is usually possible to see the terminal bud scale scars for several years. This is useful for measuring how much the twig has grown each year. Shoot growth can be an indication of the health of the tree or can be useful in determining when a tree has become stressed.

The twigs provide support for leaves, flowers, and fruits. Branches support twigs, and the trunk supports the entire crown. Yet branches ˜

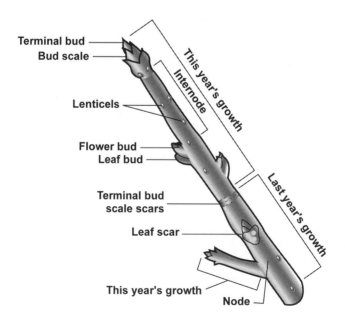

2.5. Twig anatomy showing twig extension growth.

are not simply outgrowths of the trunk. Instead, branches and trunks have a unique attachment form. It is critical to understand the relationship between a branch and its parent branch to prune a tree properly.

The annual production of layers of xylem form the conical, socket-like attachment of a branch to the trunk or parent branch from which it arises. The overlapping tissue at the junction forms a bulge around the branch base called a "branch collar." In the crotch, the branch and trunk expand against each other. As a result, the bark is pushed up, forming the branch bark ridge. If bark in the crotch becomes enclosed, grown over and around by wood, it is called included bark. Included bark weakens the branch attachment.

Each branch of the tree is similar in structure and function to the entire tree crown. Each branch is somewhat autonomous and must produce enough food to sustain itself. In addition, each branch must store its own reserves and contribute food to its parent branch, the trunk, and the root system. A branch that cannot sustain itself will draw out its reserves and decline until it eventually dies.

The "food factories" of a tree are the leaves. The leaves are displayed to the sun to absorb as much light energy as possible. The complex se-

2.6. Annual layers of overlapping xylem tissues, which form the branch collar and branch bark ridge, attach subordinate branches. Each branch must produce enough sugar to support itself and still export sugar to the supporting trunk and root system.

ries of reactions known as photosynthesis convert light energy into simple sugar molecules. The sugar may be used immediately to provide energy for growth and development, or may be stored as starch in the parenchyma cells throughout the tree.

From the Soil Down

The root systems of trees are commonly misunderstood. Many people think a tree puts down a taproot that branches into a system more or less mirroring the scaffold of the canopy. This image is far from reality. The entire root system of many mature trees can be in the upper two or three feet of soil. Most of the fine, absorbing roots are in the upper eight inches. This is why trees are so sensitive to construction damage, compaction, and changes in grade.

Roots grow where moisture and oxygen are available. Although they do not seek out moisture or porous soil, they will grow where the conditions are favorable and cease to grow where conditions are poor. The root system of a tree may extend laterally for considerable distances. Roots of trees grown in the open often extend two to three times the radius of the crown. The extent and direction of growth is more a function of environment than genetics.

The roots of trees serve four primary functions: anchorage, storage, absorption, and conduction. Larger roots are similar anatomically to the branches of the tree. Their main functions are anchorage and stor-

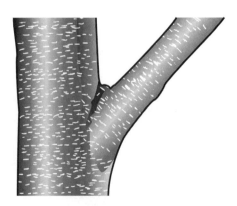

**2.7. Normal, subordinate branch development
produces a strong attachment.**

age. Although they do not play a major role in absorption, large roots will absorb water and solutes when in contact with them. The principal absorbing roots are in wide networks of short-lived, nonwoody roots.

The absorbing roots of most trees are modified into mycorrhizae. Mycorrhizae, or "fungus roots," are a symbiotic relationship between certain species of fungi and the root tissues. In symbiosis, both organisms, the tree and the fungus, benefit from the living arrangement. The mycorrhizal fungi derive nourishment from the roots of the tree. In turn, the fungi aid the roots in the absorption of water and essential minerals.

Putting It Together

Trees require a few basic materials that they draw from their environment; oxygen and carbon dioxide from the air, and water and essential minerals from the soil. Water and essential elements in the form of ions are absorbed into the roots by osmosis. Osmosis is the movement

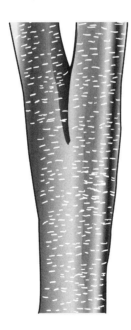

2.8. Codominant stems lack the overlapping wood development, and are weakly attached. If included bark is present, the likelihood of failure increases.

of water through a membrane from a region of high potential (water concentration) to a region of low water potential. Pure water has the highest potential; adding anything such as sugar or minerals lowers the potential. Water will normally move into roots where the water potential is lower than the surrounding soil. If the water potential is lower in the soil than in the root cells, water will actually move out of the roots into the soil. An example is when salt concentrations are high in the soil such as from excessive fertilization or deicing salts.

The tree uses some of the water it takes up for growth, photosynthesis, and other metabolic processes. A large percentage of this water, as much as 98 percent, is lost through transpiration. Transpiration is wa-

2.9. The majority of the fine, water-absorbing tree roots are in the upper eight inches of soil. Few tree roots are deeper than three feet in most landscape soils.

ter loss in the form of vapor through small openings in the leaves (stomata). This water loss dissipates heat and creates the "transpirational pull" that helps move water through the xylem.

Temperature, humidity, and available water affect the rate of transpiration. Transpirational water loss is also affected by anatomical features such as the thickness of the leaf cuticle, presence of leaf hairs, and the number and location of the stomata. Some trees are adapted to hot and dry conditions with small leaves, a thick cuticle, and sunken stomata.

Water moves throughout the tree from regions of high water potential to regions of low potential such as is created by transpiration. The photosynthetic process also creates regions of low potential by consuming water and producing solutes (sugar). Water is drawn into the leaves and into cells containing chlorophyll, the pigment that gives plants their green color and that absorbs light energy.

Photosynthesis is the process by which green plants use light energy to build sugar molecules. Literally, photosynthesis means "putting together with light." The raw materials necessary for photosynthesis are carbon dioxide and water. Light energy drives the reaction that pro-

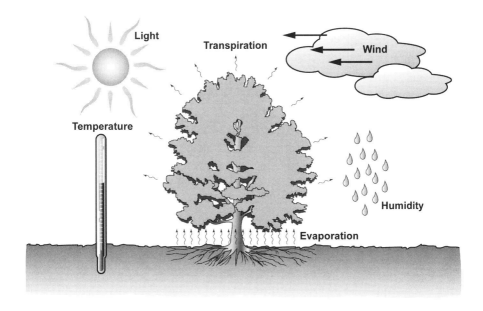

2.10. Evapotranspiration rates are affected by environmental conditions including light, temperature, wind, and humidity.

duces sugar and oxygen. The oxygen is given off into the atmosphere through the stomata.

The sugar products of photosynthesis are sometimes referred to as photosynthate. Photosynthates are the building blocks for many other compounds required by the tree. When combined with essential elements such as nitrogen, potassium, sulfur, and iron, compounds such as amino acids, proteins, and fats are produced from photosynthates. Much of the photosynthate is stored in the form of starch for later energy requirements.

Sugar is transported in the phloem from areas where it is produced to areas where it is utilized or stored. Most plant functions require energy, including phloem transport of sugar. Most of the sugar produced is used or stored close to where it was produced. Some is transported all the way down to the roots.

Respiration is the process by which the chemical energy generated by photosynthesis, and stored as starch, is used by the tree. Sugars and starch are actually chains of carbon, hydrogen, and oxygen molecules bound together. When the bonds are broken, energy is released, and carbon dioxide and water are given off. The energy released is used for biological functions.

Both plants and animals respire. Respiration is a constant process of converting food into energy. Only plants produce their own food, however. Thus it is important that overall photosynthesis exceeds respiration if growth is to occur. When respiration takes place in the absence of photosynthesis, the tree must rely on stored energy reserves. If this occurs over a long period of time, the tree will eventually run out of energy and die. A practical example is a tree that is repeatedly defoliated. Without foliage, photosynthesis stops, and the tree cannot produce and store starches. Yet as long as the tree is living, respiration takes place and energy is consumed. It must rely on stored food.

Oxygen is required for normal (aerobic) respiration. Under anaerobic conditions such as flooding, oxygen is scarce and the normal respiration reactions cannot be completed. This leads to the buildup of toxic compounds within the cells, and the tree is stressed. Prolonged oxygen deprivation will cause the tree to die.

The Communications System

Trees, like all living organisms, respond to environmental stimuli. Developmental responses to light, gravity, and temperature are essential to survival. Coordination of responses is controlled, in part, by growth regulation chemicals, or hormones. Plant hormones are naturally occurring compounds that act in small quantities to regulate growth and development. They serve as signals to trigger responses within the plant. Hormones control such things as root and shoot growth, fruit ripening, and leaf drop. The major hormone groups include auxin, gibberellins, cytokinins, ethylene, and abscisic acid. These hormones work together, maintaining a delicate balance of growth and development.

Auxin, the best-known plant hormone, has been linked with many growth responses and developmental processes. It is produced in the shoot tips and is responsible for the apical control effect already discussed. High levels of auxin, produced in the terminal bud, inhibit the growth and development of lateral buds and shoots. Auxin is also important in root development. In fact, synthetic auxins are often used by plant propagators for rooting cuttings.

Another group of hormones is cytokinins, which are produced in the roots. Cytokinins are known to be critical in shoot initiation. Auxin and cytokinin interact in the tree to affect the timing of root and shoot growth. Two examples, although oversimplified, illustrate the delicate balance between auxin and cytokinin, and the roles they play.

When a tree is topped, most of the shoot tips are cut off. The auxin source is effectively removed, and apical control is eliminated. Cytokinin levels become high compared to auxin levels. As a consequence, dormant buds below the cuts are triggered to develop. Multiple shoots are produced below each cut throughout the tree's crown. This growth response is a survival mechanism to produce as many leaf-bearing shoots as quickly as possible.

Inversely, when most of the roots are removed, such as in transplanting, the tree's source of cytokinins is reduced dramatically. Auxin levels from the shoots remain high in relation to cytokinins, and the tree is signaled to begin rapid root production. This is one reason why pruning living branches at the time of planting should be kept to a minimum.

The Defense Department

A developmental process unique to trees is the ability to compartmentalize decay. Compartmentalization is a process by which trees limit the spread of discoloration and decay. After a tree has been wounded, reactions are triggered which cause the tree to form boundaries around the wounded area.

Biochemical and physical boundaries are established in vessels, old xylem cells, and ray cells. At times these boundaries are insufficient to prevent the spread of decay inside the tree. The strongest boundary is established by the new xylem formed after the wounding. The tree may decay inside to the point of becoming hollow, yet the new wood is normally not invaded.

An old tree surgery practice was to "clean out" cavities and fill them with various materials. However, the cleaning process was likely to break the boundary of new wood and introduce the decay-causing organisms. Current recommendations are to avoid this technique and allow the tree's defense system to wall off the decay.

A Balanced Budget

Sometimes the growth and development of trees is better understood in terms of an economic analogy. A tree's sole source of income (food) is the sugar that is produced in photosynthesis. Limiting factors include the number and shape of the leaves, the available sunlight, water, and carbon dioxide, and adequate supplies of a few essential minerals. Each

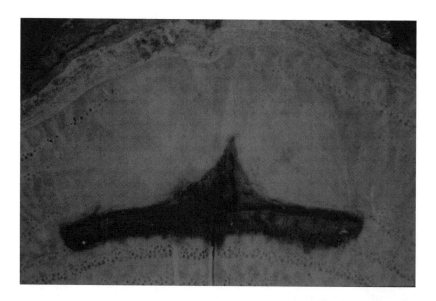

2.11. Wounds are compartmentalized by chemicals produced in existing wood and by the new wood produced following wounding.

branch must sustain itself, but also must pay "taxes" in the form of sugar contributions to parent branches, the trunk, and the roots.

Health problems such as nutrient deficiencies, drought, and defoliating pests can dramatically reduce the income potential. In such cases, the tree must draw out of savings (stored starch reserves) enough money to support itself until more income can be generated.

When a branch is pruned, most of what is removed is the stored starch for that branch, but each branch pruned further reduces the "tax base" for the rest of the tree. Furthermore, some reserves will be required to close and defend the wound. Young trees have a high income to expense ratio and are very tolerant of pruning. In addition, wounds are relatively small and easily closed.

As trees mature, they grow larger in bulk. Although they have a large savings capacity, their income to expense ratio decreases with each pass-

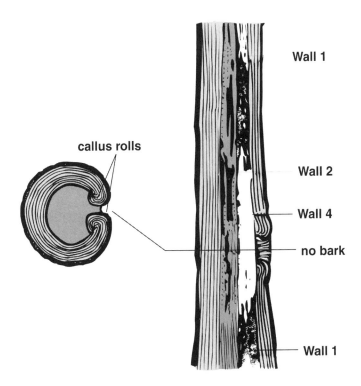

2.12. Compartmentalization of decay. Wall 4 prevents decay from entering new wood. Wall 3, not shown, and Wall 2 have failed to prevent decay from spreading laterally and internally.

ing year. Very large, mature trees grow little in height and spread, and put out approximately the same leaf volume each year. They are, at this point, on a "fixed income." Nevertheless, their tax rates do not decrease. Health stresses can be very serious for these senior citizens that lack the vitality to mount strong defenses against disease and decay. Removal of branches at this point reduces the sorely needed income potential of the tree. If a large wound is created, it will probably never close and may lead to significant decay if the tree lacks the defense capability to compartmentalize.

If tree caretakers have a better appreciation of how trees make a living they can make informed decisions regarding tree maintenance. All arboricultural practices have an impact on the health of the tree. Some, such as pruning, cabling, and the installation of lightning protection, wound the tree. The benefits must be weighed against the injuries. Other treatments, such as fertilization, can be beneficial when required, but harmful when administered to excess. It is essential to understand tree biology before attempting to diagnose problems and recommend a course of action.

Summary

1. It is essential to understand the biology of a tree in order to manage it in a way that supports its growth and development. Arborists use knowledge of tree anatomy and physiology to diagnose problems, assess potential, and prescribe treatments.
2. The anatomy and physiology of trees is much different from that of turf. With the exception of palms, trees are dicots, producing secondary meristems and woody tissues. The cambium, located just inside the bark, produces xylem tissues to the inside and phloem to the outside. The xylem carries water and essential minerals up from the roots and throughout the tree. The phloem transports sugar from the leaves to where it is used or stored.
3. Trees grow from meristematic tissues, or areas of rapid cell division. Apical meristems are located in shoot tips and root tips, and allow elongation of the roots and shoots. The cambium is a secondary meristem, allowing growth in girth.
4. The leaves are the food factories of the tree. Through a process called photosynthesis, they produce sugar. The sugar is usually

utilized or stored in close proximity to where it is manufactured. It is stored in nonwood cells known as parenchyma, or transported through the phloem to the roots.

5. A tree's root system is not a mirror image of the top. Most of the fine, nonwoody roots, which absorb water, are located in the top 6–8 inches of soil. The roots of a tree may extend far from the trunk, often several times the radius of the spread of the crown.

6. Trees require a few basic materials that they draw from the environment; oxygen and carbon dioxide from the air (above- and below ground), and water and essential minerals from the soil. If the tree cannot get sufficient quantities of any of these, it will be stressed, and may decline or die.

7. Everything that happens in a tree's environment has the potential to affect the health of the tree. By understanding how trees grow and develop, managers can be better equipped to make informed decision regarding tree maintenance.

3

Trees vs. Turf

Why Trees and Turf Don't Get Along Well

Trees and turf tend to be mutually exclusive in nature. There are some exceptions such as the oak savannas where frequent fires limit the establishment of new trees. But for the most part, you won't see many trees growing in the prairies or grasslands, and grass is not common on the forest floor.

In nature, the rule seems to be whichever establishes first has the advantage. A forest or stand of trees produces too much shade for grass to grow. In open land the roots of grasses are more aggressive, faster to colonize, and establish a dense root system that excludes trees.

The landscape of the golf course is an unnatural ecosystem. Man forces two somewhat incompatible plant types together and expects optimum performance from each. There are, of course, some commonalities—trees and turf both need sunlight, water, and the same basic nutrients. The problem is, these shared needs lead to competition, and to compound the problem, the maintenance needs for each can be detrimental to the other.

The competition between trees and turf is fought on several fronts. They compete for water and nutrients. They compete for sunlight. Their root systems vie for space underground. The battle is not always a passive contest: each has ways of inhibiting the growth of the other.

Competition

Much of the competition takes place below ground. There is a misconception that the roots of a tree are much deeper than the grass roots. Although some tree roots can be found well below the surface, the fine roots that absorb water and minerals are generally in the upper few inches of soil. Roots of all plants need oxygen to survive, grow, and function. Although roots do not "seek out" resources, they do grow where conditions are optimal. Since aeration, nitrogen, and moisture availability is usually best near the surface, the roots of the trees, turf, and any other landscape plants will share this space. This is why surface roots can be a problem with many species of trees.

Any given volume of soil can support a finite root population. The number of roots will vary with the soil type and conditions. Since turf roots tend to respond quicker to resource availability and to colonize faster, they usually have the advantage. Numerous studies have shown that turf roots will take up the majority of nitrogen fertilizer applications when sharing soil space with tree roots.

Trees, on the other hand, have tremendous absorptive power allowing them to draw large quantities of water from the soil. On a hot, summer day a stand of mature trees can easily pull hundreds of gallons of water from the surrounding soil. Competition for soil moisture can lead to stress in both the trees and the turf. Stressed plants may be predisposed to other problems with pests and diseases. While the trees and turf are struggling to get by, weeds with low moisture requirements are likely to become established.

Shade

From a golf course superintendent's point of view, shade is the root of all evil. Shade from trees contributes directly or indirectly to many of the turf density and health problems on greens, tees, and fairways. Shade reduces the quantity of light available to turf and the length of time it is available. Shade leads to reduced stand density, increased root competition, and increased weed invasion. The trees reduce air circulation, which in turn causes additional problems with turf diseases. Shade is a major stress factor for turf.

There are actually three aspects of shade that affect the light available to turf: light quantity, light quality, and duration. The quantity of light that penetrates the tree canopy to the turf will depend on the height and spread of the trees, the distance from the turf, the location in relation to the sun's direction (which in turn varies by season and time of day), the tree species, and the density of the trees' canopy. Obviously, tall trees cast long shadows, and the closer they are to the turf, the less light will be available to the grass. Trees that grow on the south and east sides of the turf tend to be more troublesome because they block the morning sun, which is critical to turf growth. Trees with dense canopies block more sun than trees with open canopies.

The quality of light available is also affected by the shade of trees. Tree leaves absorb the violet/blue and the orange/red wavelengths of light from the spectrum. The remaining light that passes through is rich in the green/yellow wavelengths, sometimes called "green shade." It is the blue wavelength that is important for turf growth. In effect, trees

3.1. Shade can be the biggest problem in trying to maintain trees and turf together. Shade must be considered in planning and management of golf course trees.

"mine" the sunlight as it passes through their canopies, leaving the less photosynthetically-active wavelengths in the light that reaches the ground below. This depletion of red light alters growth processes. More energy is allocated to leaf blade growth than root and rhizome production, causing the turf to produce longer, thinner leaf blades and poorer root development.

The duration of light reaching the turfgrass depends on several factors. The sun rises and sets at different angles from the horizon depending on the time of year and the geographic location of the golf course. These factors also affect the length of daylight. The size and location of the trees in comparison to the angle of the sun affect the duration of sunlight to the turf.

Reduced turf density opens the door to invasion by shade-loving weeds such as wild violet, ground ivy, and nimble. These weeds can be difficult to control, especially since the grass may not being filling in enough to prevent weed reestablishment. Low light levels and high humidity create conditions favorable to diseases. Turf diseases associated with shade include powdery mildew and dollar spot. In addition, moss and algae may thrive in shady environments.

3.2. Shade management requires analysis of sun direction and angles, as well as tree species, height, spread, and location in relation to greens.

There are some varieties of turf that are somewhat shade tolerant, and this may be a partial solution in some limited circumstances. Shade tolerant grasses tend to be less tolerant of wear. Golf courses require high density stands that can be mowed to a low height, yet withstand high levels of traffic. Many of the shade tolerant turf varieties cannot meet these standards.

Growth Inhibition

In addition to competing for light, water, and nutrients, trees and turf can inhibit the growth of other plants by releasing chemicals that inhibit germination or development. This phenomenon is called allelopathy. Although there is still much to learn about the chemical effects that plants have on one another and on insects and diseases, there is a body of research documenting suppression of one species by another.

Allelochemicals produced by plants may be in the form of terpenes, phenolic compounds, organic acids, tannins, steroids, and other compounds. They may be leached from the roots into the soil, volatilized into the air, exuded from plant parts or released in the decaying process of

3.3. Allelopathy is the inhibition of the growth and development of one plant by another. Here a walnut tree's production of juglone affects the white pines growing nearby.

dead tissues. The best-known example of allelopathy is the inhibition of growth of various plants by juglone, a chemical produced by walnut trees.

Many trees are known to have allelopathic effects on the germination or development of other plants. Documented cases include species of pine, planetree, maple, hackberry, eucalyptus, sumac, and others. In fact, allelopathy may play an important role in the natural succession of trees from pioneer to climax species. Not much is known, however, to what degree chemical inhibition limits turf growth and establishment within the root zones of trees.

There is also evidence that grasses can inhibit the germination and development of young trees, especially conifers. Research shows reduced growth of trees that are surrounded by turf as compared to trees that have mulch applied over the root system. The question is how much the phenomenon can be extrapolated to larger, established trees. Trees that are growing in a mulched environment do exhibit better growth and less stress than those growing together with turf. This can no doubt be attributed to competition and other factors in addition to any allelochemical effects.

Turf growing under a tree's canopy also suffers a range of physiological stresses. Turf will have reduced vigor and reduced tolerance to drought, heat, cold, and wear. Turf affected by trees tends to have shallower roots, reduced stand density, and decreased food reserves. These stresses predispose the turf to diseases, weed invasion, and pest problems. In addition, the capacity of the grass to recuperate from traffic and wear is significantly diminished.

There are many indirect effects of trees on turf and turf on trees. Even the way golfers play can affect the tree/turf interaction. For example, if a tree's branches hang over one side of a fairway, the players may all elect to tee up on the opposite side of the tee box. This in turn can cause uneven wear and stress on the turf in the tee area. Conversely, golfers may gather or wait their turns under the cooling shade of a tree's canopy, causing more wear in that area.

Some of the strongest influences in tree/turf relationships result from maintenance practices. Tree care procedures such as pruning and removal can be very injurious to turf. Turf maintenance such as mowing, string trimming, herbicide applications, and irrigation use can hurt trees. The challenge is for the superintendent to recognize the problems involved, and institute new methods, or modify old techniques, in ways that allow a more healthy coexistence between trees and turf.

Tree Care in Relation to Turf

As already discussed, a major turf problem created by trees is the limited light. Grass requires the critical morning sunlight to optimize growth, health, and stand density. In addition, if shade can be reduced, turf will come back earlier in the spring, making the course playable sooner. In northern climates, direct sunlight hastens snow and ice melt. This reduces some disease problems and gives the turf a longer growing season. It also expedites spring cleanup.

Managing Shade

Shade is best managed before it becomes a problem through careful planning of tree placement in relation to turf and through the selection of appropriate tree species. Unfortunately, golf course superintendents are usually facing the problem long after the planning and selection stages. This often leaves a choice of two management decisions, pruning or removal. Actually in most cases the answer is a combination of both, but neither may be popular with the greens committee or management board.

There are several methods of deciding what has to go to provide more sunlight to the turf. Most frequently the decision is made based on the superintendent's gut instinct and experience with the course. Another option is to use time-lapse photography to record shade patterns at different times during the day and different times during the year. While this is more precise than the "cut and see" method, it can take a long time to make a decision.

A newer, more sophisticated technique is to use a sun location system called the SunSeeker™. This computer-aided system plots the sun's coordinates at 15-minute intervals and uses that information to record sunlight and shade patterns for different times of the year. The information gathered is then used in the decision-making process to determine which trees should be removed, which trees should be pruned, and to some extent, the specific limbs that should be removed or pruned. It can also be used to determine where to plant trees and the distance from the greens. Using this kind of modern technology not only helps the golf course superintendent to conserve as many trees as practical, but can help convince the management of that goal.

The pruning and removal decisions should involve the expertise of an arborist. Arborists can help predict the growth responses of the trees following selective thinning and pruning. An important axiom to re-member is that trees will grow into the voids created. The directed growth can be more aesthetically appealing, structurally sound, and healthier for the trees if the cuts are made in the right places.

If a stand of trees must be thinned out, an arborist can help identify the trees that do a disservice to the course. A good knowledge of various tree species' attributes and characteristics is invaluable at this point. Consider keeping the trees with more open canopies, less debris-drop-ping tendencies, and few insect and disease problems. Keep in mind the importance of maintaining species diversity.

When pruning trees, there is often a tendency to remove too much. Topping should not even be considered. Thinning is preferred. The rule of thumb is not to remove more than one-fourth of the tree's foliage-bearing crown in a single pruning. Even this is too much for many large, mature trees. If the crown is thinned too much, it will be stressed, and will probably produce many watersprouts (suckers) along its branches to compensate for lost foliage. This defeats your purpose of crown thin-ning to allow more light penetration.

3.4. Trees should be far enough away from greens to allow the crucial morning sunlight to penetrate.

It may help to "raise" the trees' crowns to improve light penetration and duration and air circulation around the turf. Crown raising involves the removal of lower branches on the trees. Most tree species are quite tolerant of this pruning practice within limits. Removing the lower branches reduces the formation of good trunk taper, which is important both for the tree's health and its structural stability. Raising the crown too much may increase the likelihood of windthrow and will probably induce watersprout production. Since trees tend to fill in the voids opened up by tree removal and pruning, this must be taken into account when deciding how much and how often to thin.

Even the terminology that you use can be important when it comes to tree removal and pruning. Given the reluctance of greens committees to approve the removal of any trees, you must demonstrate the necessity of reducing shade while confirming the value of the trees and your desire to preserve them. When discussing groups or stands of trees that require some removals, it may be preferable to recommend "selective thinning." This implies a thought process that may not be as clear if you say you must cut down some trees. By the same token, "selective prun-

3.5. Raising the canopies of trees can allow sunlight penetration. If the pruning is too severe, however, it can lead to health stress or tree failure from windthrow.

ing" sounds less drastic and better planned than tree trimming. There is more detailed information about pruning in Chapter 7.

Root Control

It is not unusual to read planning specifications that call for deeply rooted trees species. The concept that some trees have inherently deep roots while others have shallow root systems is somewhat misguided. Trees' root growth is more a function of environment than genetics. Roots will grow where conditions are best for root growth. In most cases, that is near the soil surface. Most of the fine, absorbing roots will be found in the uppermost foot or even six inches of soil.

When tree roots grow and expand they can create surface roots. Surface roots can be a major problem on golf courses. Besides ruining the appearance of the turf, they can interfere with mowing equipment, interfere with the play of the game, and can even become a safety hazard. Tree roots are absolutely unacceptable on greens and tees.

Golf course superintendents always want to know to what extent they can prune or remove tree roots without bringing about the demise

3.6. This tree failed as a result of root pruning for turf renovation.

of the tree. There is no simple answer to this question, as it varies with species, age, and conditions. There are a few rules of thumb.

The first thing that must be well understood is how trees respond to root pruning. Assuming the cuts are relatively clean and do not cause the root to die back, most roots will sprout new growth from just behind the point of the cut. There are usually anywhere from one to six new roots created from the old cut root. Generally, all of the new roots that develop will grow in the same direction as the original root. If you are root pruning for the purpose of transplanting the tree, this is good. If you are trying to stop root invasion onto a golf course green, this is bad.

As mentioned, the amount of root pruning that a tree will tolerate varies. Most vigorous, healthy trees can tolerate the onetime loss of 25

3.7. Leaf litter can be a headache for superintendents. It often requires blowing the leaves off greens and tees daily.

percent of the root system. Note that if major support roots are severed there may be an increased risk of the tree blowing over. As a rule, never prune more than 33 percent of the roots under a tree's canopy. Older trees, however, may not tolerate this treatment. Also, root pruning should not be performed more frequently than every other year. If roots must be pruned, avoid doing it during periods of active shoot growth or stressful periods such as hot, dry summers that are not favorable for root growth.

Another option may be to use root barriers. Root barriers are installed vertically in the ground to prevent root penetration and redirect root growth. There are a number of products available that use various strategies to redirect the growth of roots. Studies show, however, that roots tend to grow down and underneath the barriers, then upward at an angle toward the surface again. This has the effect of buying you a few feet just beyond the barrier where roots will not interfere. Another problem is that roots can grow over the root barriers if the barriers are not at least flush with grade. This, however, could interfere with play. Thus these barriers may be helpful for protecting the integrity of a cart path, but they probably aren't the solution for greens.

Fertilization

There is a long-standing, but inaccurate, belief that trees must be "deep root" fertilized. This belief is associated with the myth that a tree's root system is an underground mirror of the crown. Since most of the roots that have the ability to take up minerals are actually in the upper few inches of soil, it makes little sense to place the fertilizer deeper.

Where tree roots and grass roots occupy the same soil volume it is impossible to fertilize the turf without also fertilizing the trees. Studies have shown that in this scenario, the turf roots absorb a much larger percentage of the fertilizer than do the tree roots. The turf may have up to six times the root mass per given soil volume as the trees, and they respond more quickly to changes in soil fertility. The question is, do the trees require supplemental fertilization?

The primary nutrient that may be in limited quantity for both turf and trees is nitrogen. Golf courses usually provide plenty of nitrogen to the turf to maintain the high standards of color, density, and stress tolerance. Where tree roots are benefiting from this windfall supply of nitrogen, additional application may not be required. Trees that are not near the turf may require additional fertilization. Others may need an application for other minerals, especially certain micronutrients. An example might be pin oaks that are suffering from chlorosis due to inadequate uptake of iron. The key to any fertilization program is to base the application on the plant's needs. Soil and foliar analyses can provide the information required to make an educated decision about nutrient needs. Remember: treatment without diagnosis is malpractice.

Arborists often use fertilizer application techniques such as soil injection or the drilled hole method that place the fertilizer below the soil surface. This may be beneficial in that it reaches some tree roots that are below the surface and there is less loss to the turf, but it can also miss the majority of the absorbing roots. Soil injection and drilled hole fertilizer application can also cause dark, vigorous patches of grass in the vicinity of the application holes. If the trees are located away from the turf, the application technique is less critical. In turf, surface application is still the least expensive and most practical. But in turf, the application is less likely to be required at all.

There is a mounting body of evidence suggesting that excess fertilization can predispose trees to other problems, particularly insect problems. Researchers have found that triggering trees to produce abundant vegetative growth may cause the tree to direct fewer reserves to defense and storage. Furthermore, the lush, new leaves may attract certain foliage-feeding pests, while the allelopathic chemicals that deter these pests may be reduced. With the high rates of fertilizer application used by golf courses, this can be a concern even if the trees are not receiving supplemental applications.

Debris Control and Disposal

All things considered, trees require much less maintenance on a golf course than turf does. The problem is that trees create a lot of "stuff" that has to be removed. Trees shed flower parts and petals in the spring, fruit and seeds in the summer, and leaves in the fall. Some trees are constantly dropping twigs and branches or pieces of bark. To the golf course superintendent, this means blowing debris off greens and tees every day before mowing. And compounding the problem, daylight hours are precious in the autumn.

Leaves can accumulate in ponds, lakes, and streams along the course, upsetting the ecobalance of the water or clogging pumps. With thousands of trees on or around a golf course, the volume of leaves that pile up is staggering. When collected and cleaned up, the superintendent has the problem of disposal.

Where practical, it is good to keep the leaves and other organic debris around the trees. The exception is where certain diseases must be controlled through good sanitation practices that include leaf removal. Leaves and organic debris are natural sources of essential minerals. They break down, conditioning the soil and adding organic matter. As decomposition takes place the soil is enriched, conditions are improved for mycorrhizal activity, and a natural mulch is created. The need for supplemental fertilization can be reduced or eliminated.

In addition to the natural cycle of falling leaves and other plant parts, a tremendous amount of debris is created when trees are pruned or removed. Most tree care companies and many golf courses own chippers that are used to grind branches into wood chips. The wood chips created can be valuable as mulch. Most courses that chip their own brush generate a quantity of chips that they can easily use around the course. If a tree service is hired, the superintendent and arborist must decide

3.8. When working on golf courses, arborists must take extra precautions to avoid damaging turf.

what is to be done with the chips. It may be necessary for the tree service to haul away excess chips.

Another alternative in some cases is to burn debris collected from tree pruning and removal. This may not be an option in many jurisdictions due to laws that limit burning in urban areas. Also, if the volume is large, burning may not be the most practical choice. The smoke is a pollutant, an irritant, and sometimes a public relations problem.

Protecting Turf

Trees are the largest of all plants, and when they are pruned, the "clippings" can weigh hundreds of pounds. Tree maintenance can be damaging to turf. Special precautions must be taken to limit the damage that occurs.

One way to limit turf damage is to carefully plan the timing of the maintenance. Many superintendents plan the major tree maintenance and removal for the off-season. In northern climates, this means the winter when, with luck, the ground is frozen. In southern climates, the opposite may be the case. Tree work may be done in the heat of summer when the ground is hard and dry. Some courses may even reduce irrigation and allow the turf to go dormant.

When tree work must be performed at other times during the year, the biggest consideration is usually the equipment. Trucks and chippers weigh thousands of pounds and must be kept off the fairways as much as possible. Some companies that do a great deal of golf course tree maintenance have started using high flotation tires and/or turf tires on their equipment. This can make a tremendous difference in reducing the impact on the turf.

It may be necessary, in some circumstances, to keep the trucks and chippers a good distance away from the area where the tree work is taking place. Although some companies prefer to do the majority of their tree pruning using aerial lift trucks, there are few trees that cannot be pruned by climbing. In fact, the quality of the work is usually far superior if the trees are pruned by climbing as opposed to using a bucket truck.

Branches must be handled more carefully as they are being cut and removed from the trees. In some cases everything cut may have to be roped down to avoid causing ruts in the turf or hitting irrigation components below the ground. A tree that may be a simple "drop" removal in another setting can be a complex rigging situation on a golf course. Although the expense in labor can be considerably higher, damage to the turf is usually unacceptable, especially on tees, greens, and fairways.

The wood and brush that is created in pruning or removing trees must be moved to where the trucks and chippers are parked. Brush should not be dragged across golf course turf because the potential for damage is too great. It will probably be necessary to cut the brush into much smaller sections than normal. It can then be carried or moved by a golf maintenance vehicle or other vehicle that is less damaging to the turf.

3.9. Young, newly planted trees often fall victim to "lawn mower blight," especially if not protected by a mulched area.

Plant Health Care and the Environment

Golf courses can be a sanctuary for wildlife in the midst of urban expansion. The trees are a habitat for many species of birds and small mammals, providing shelter and food. Superintendents and arborists must take this into account when making decisions about tree management. When it is practical and hazard potential is very low, consider leaving dead tree trunks and snags in place. They are nesting sites for a number of animals. The animals, especially certain birds, can be valuable by reducing mosquitoes and other insect pests. Wildlife enhances the natural environment of the course, and people like to think of golf courses as natural green spaces. Of course, safety concerns override the desire to maintain a habitat, a fact that usually restricts leaving snags only in isolated, low-traffic areas.

A critical element that is part of maintaining an ecologically balanced environment is the principle of plant health care (PHC). A good PHC program emphasizes the need for sensible planning. Maintaining trees, turf, and other plants in good health requires more than fertilizing and pest control. It involves considering each plant or group of plants in relation to the other plants (and animals) in the landscape. Plant health care is a matter of looking at the big picture, considering the basic needs of each plant, and attempting to optimize the growing conditions for each without sacrificing or harming the others.

Maintaining plant health is integral in reducing the need for intervention in the form of pest and disease control. The days of broadcast spraying of pesticides on trees are gone. In the "old days," massive doses of chemicals were applied, knocking down populations of pests, beneficial insects, and sometimes causing collateral damage to nontargeted plants and animals. Now the goal is to monitor plant health and intercede only when necessary to maintain the desired level of plant health and aesthetics.

Turf Care in Relation to Trees

The primary concern of all golf course superintendents on a daily basis is maintaining the turf. This emphasis is understandable and justified. The turf is critical to the game of golf. It must look great, it must be healthy, and it must be meticulously maintained to facilitate play. The grass requires much more frequent attention than the trees or other landscape plants require. Unfortunately, the turf is often maintained at the expense of the trees, and in some instances, to their detriment.

Mowing

Golf course grass is mowed much more frequently and to a lower height than grass in most any other setting. Greens are generally mowed every day. The problem is that the mowing equipment can damage trees. Each time a mower is bumped against the trunk, or the worker uses the tree to pivot around, the trees are injured. The injuries grow progressively bigger and deeper since the equipment hits in pretty much the same places over and over again. Eventually the damage progresses internally through the phloem, cambium, and xylem of the tree. In the worst-case scenario, the trees are eventually girdled and die. Those that are not killed will be stressed. The wounds may serve as entry points for diseases, borers, or other insects.

Mowing damage is of particular concern in the spring before new tissues on the trunks of trees have hardened off. When the cambium is rapidly

3.10. Injuries from mowers and string trimmers can cut off vascular tissues under the bark, and can provide an entry point for decay organisms.

dividing, producing new phloem and xylem, the bark is said to be "slipping." Trees are much more vulnerable to physical injury at this time.

Mowers are not the only culprits. Spray rigs, aeration devices, and any other equipment used around trees can cause mechanical injury. One of the worst offenders is the notorious string trimmer. Workers don't realize the degree of damage that can be caused by the whipping action of a nylon string. A tree's bark can only provide so much protection against these devices. Young, thin-barked trees can be damaged almost immediately. Golf course maintenance workers must be educated about protecting and caring for trees. One of the best ways to eliminate or reduce this problem is to simply mulch around the trees.

Irrigation Concerns

Irrigation is a double-edged sword. Supplemental irrigation when and where it is needed can, in the long run, save the life of a tree. Too much irrigation, or irrigation at the wrong time can be stressful, even leading

3.11. Irrigation requirements are less in shaded areas than in the full sun. Excess irrigation water can accumulate, causing stress to nearby trees.

to the eventual death of a tree. Irrigation without proper drainage is likely to be a problem.

The water needs of trees vary with species, soil type, and environmental conditions. Some trees take a big gulp of water in the spring and their water use throughout the summer levels off somewhat. Others maintain a very high level of water absorption throughout the growing season. Native trees will be adapted to normal environmental conditions for that region. Thus, irrigation of golf courses in arid climates can be a problem for native trees.

Frequent, shallow watering can encourage surface root formation. Excess irrigation of trees can lead to root crown decay and Armillaria root rot. It can cause poor root health that is manifested as nutrient deficiency, decline in the canopy, poor growth, or a host of other symptoms. Stressed trees will be subject to other problems including insect activity and poor defense against decay.

A good irrigation management plan will include an assessment of all of the landscape plants, not just the turf. Head placement, type, and spray angles should be set according to water requirements and drainage patterns. Keep in mind that shaded areas will require less supple-

3.12. When the soil is saturated, strong winds often blow trees over, since the root system has less anchorage. Note how shallow the root system is on this mature tree.

mental irrigation than areas in the full sun. Avoid allowing irrigation spray to hit tree trunks directly. Monitor soil moisture levels and adjust irrigation accordingly.

Aeration

It seems logical to think that aeration of turfgrass will benefit trees that share the same root zone. This is not always the case, however. Hollow tine aeration tends to benefit the turf roots more than the tree roots, since turf roots respond more quickly to the improved aeration. In fact, hollow tine aeration can lead to the elimination of tree roots in the upper few inches of soil due to improved turf root growth.

Chemical Treatments

Most golf courses utilize a large spectrum of chemicals to maintain the turf at such a high level under difficult conditions. Herbicides, especially broadleaf weed killers, are an important part of that arsenal. It is

3.13. Irrigation should not be allowed to spray tree trunks and canopies. This can result in crown rot and various fungal diseases on the foliage.

important to remember, however, that most trees are broad-leaved plants and can be injured or killed if high enough doses reach them. Growth regulator herbicides such as 2,4-D and dicamba are systemic, which means they move throughout the plant. Exposure may be the result of drift, accidental spraying or movement in the soil. Turf managers must keep in mind that "weed and feed" fertilizers contain herbicides, which can damage trees.

Symptoms of herbicide damage to trees vary somewhat depending on the tree species and the chemical. Some symptoms include twisted or cupped foliage, and twisted shoot tips. New foliage may exhibit parallel venation or interveinal chlorosis. The leaves may all appear wilted. The

3.14. Trees close to greens often decline or die shortly after the greens are renovated. Increases in grade and cutting of tree roots are two causes.

foliage may turn yellow or may show marginal chlorosis or necrosis (dead tissues).

The extent of tree injury usually is not readily apparent. Most often, damage is not permanent, but young trees have been killed outright from accidental exposure. To minimize the chances of injuring trees, always read and follow label directions. Labels detail all restrictions and may warn of possible phytotoxity to certain groups of plants. Apply on cool, calm days if possible, to reduce volatilization and drift.

Greens and Tees

The special attention required to maintain greens and tees tends to amplify the tree problems associated with turf management simply due to the intensity. There are some extra considerations associated with both greens and tees. In some areas, fans are used to improve air circulation around turf where there are trees. This helps reduce humidity, and thus reduces disease problems. The fans are not usually a major problem for trees as long as the placement does not dry out foliage.

A problem that develops near tees is increases in grade. The turf on tees takes a great deal of abuse. Although the markers are moved constantly to give areas a chance to recover, it is common for the entire tee area to become badly damaged. Superintendents may restore the tees by adding soil and raising the entire area. This can be devastating to nearby trees. Even slight increases in grade can smother the roots below ground. The construction process of rebuilding the tees can compound the problem by compacting the surrounding soil. Superintendents must be cognizant that the process can injure and stress trees. Where practical, measures can be taken to limit the damage. If there is no workable alternative to raising the tee boxes, reduce the damage by keeping equipment away from the trees. Avoid increasing the grade any more frequently than every three years. Provide adequate aeration and irrigation to the trees and monitor their health.

Achieving a Balance

Despite the differences between trees and turf, and all of the difficulties associated with competition, growth inhibition, and conflicting maintenance needs, it is possible to achieve a balance. Trees and turf can peacefully coexist and even thrive together on a golf course. Armed with an understanding of how each affects the other, the golf course superintendent can modify the environment and the maintenance procedures to optimize the growing conditions for both.

Planning and Selection

If you have the luxury of being able to plan what trees are planted and where, then you are ahead from the start. With good planning and selection, the trees will be in less conflict with the turf as both mature. Chapter 4 of this book deals with design and construction issues and goes into more depth about good design practices. There are a few considerations, however, that deal directly with issues of tree and turf compatibility.

Because the morning sun is critical to turf health and growth, plant trees farther away from the turf on the south and east sides. This is especially important around greens and tees. Consider the mature heights and spreads of the trees and the shadows they will eventually cast. Minimizing shade can reduce turf health problems down the road.

3.15. Sprinkler irrigation can wash soil from the surface around trees, exposing roots. Mulch can protect the soil and can eliminate the need for mowing close to tree trunks.

The light intensity below a tree's canopy may be directly related to the tree's own shade tolerance. The most shade tolerant trees tend to be forest climax species such as sugar maple. These species are capable of establishing themselves in the understory of the forest. Their leaves tend to be efficient in low light intensities. At maturity, they develop dense canopies, which limit the quality and quantity of light that penetrates. As little as 5 to 10 percent of full sunlight may penetrate the canopy.

Conversely, shade intolerant species are usually pioneer species. These are trees that require open, sunny conditions to establish. They tend to be fast-growing and short-lived. They usually have open canopies, allowing as much as 33 percent light penetration. Open canopied trees such as this may work well as perimeter species. The light penetration may be sufficient for some shade tolerant grasses. With a little pruning, even full sun grasses may be able to survive in the partial shade of these species.

Maintenance

The fact that the maintenance procedures for trees and turf impact the growth and health of each other has already been established. The areas of conflict have been described and suggestions for modifications have been discussed. There are a few additional recommendations that can help avoid or reduce some of the potential problems associated with maintaining trees and turf on a golf course.

Where practical, maintain trees in groups. Individual trees are more likely to be injured by equipment, people, or errant golf balls and have a higher probability of being struck by lightning. They have a higher tendency to be surrounded by turf and will suffer more of the competition and allelopathic problems discussed earlier. Individual trees are also more difficult to work around when maintaining the turf.

The final maintenance recommendation is perhaps the most important: mulch. Mulching the root areas of the trees is probably the simplest but most beneficial thing you can do to enhance tree health and minimize competition with turf. Mulch helps retain soil moisture. It moderates soil temperature extremes, reducing root stress. It reduces competition from weeds and turf. A good organic mulch can help condition the soil and improve microbial activity.

Mulch reduces the need for irrigation. It removes the need for mowing equipment to get close to the trees, and reduces the chances of herbicide injury to the trees. Mulch is also aesthetically pleasing. It ties the landscape together and is a means of visually and physically grouping the trees together. It can be used to define the fairways or delineate out-of-bounds zones.

Apply the mulch about three to four inches deep, but do not pile it against the tree trunks. As far as the trees are concerned, the bigger the mulched area the better. Mulch groups of trees together and extend the mulched areas as far out as practical for the design of the course.

The choice of mulch is important both for the beneficial effects on tree health as well as the play of the game. Organic mulches such as pine bark, hardwood or cedar wood chips, pine straw, or others help improve the soil in addition to many of the other advantages. Finer, softer, less coarse mulches tend to work best along fairways. Golfers can find their golf balls more easily and can usually play the lie unimpeded. Pine straw is widely used too as a mulch along the fairways of some of the finest courses in the southern United States.

When and Where to Compromise

As important and valuable as trees are to a golf course, they will always come second to the turf. You really can't play golf without turf. As

3.16. Mulching around trees can reduce weed and turf competition, and protect trees from turf maintenance equipment.

a result, most of the compromises and modifications will favor the turf. Trees will always require pruning. Some will have to be removed. Maintenance on the trees will have to be worked around the golfers, the turf maintenance needs, and protection of the turf itself.

That said, however, ask any golf course superintendent how easy it is to get approval to remove trees on the course, and the superintendent will sigh in frustration. Trees are almost sacred to many greens committees. Management is reluctant to remove any trees without putting up a fight. Greens committees will ask the superintendent, "Why don't you just plant a shade tolerant variety of grass?" or "Can't you just raise the mower height a little for the turf closer to the trees?" These may be partial solutions under specific circumstances but are rarely the complete answer. Shade tolerant turfgrasses tend to be intolerant of wear. The height of the grass is usually dictated by the design of the course. Besides, it can be impractical to mow at many different heights in the same maintenance rotation.

This is where the help of a good arboricultural consultant can be invaluable. The independent view of someone who specializes in the care and preservation of trees can help you build a case. You can work together to decide which trees need to be removed and which can be pruned. The consultant can explain why these decisions were made and how their implementation can impact the course and improve turf conditions. Sometimes demonstrating a dramatic improvement in just one area can turn the tide of opinion favorably.

Trees are a very important element of a golf course. Even though they are incompatible with turf in many ways, a good management approach can help you to maintain both at a high functional and aesthetic level. The key is understanding the biological differences and basing your management program on maximizing growing conditions while minimizing conflicts.

Summary

1. Trees and turf tend to be mutually exclusive in nature. A golf course is an unnatural ecosystem in which man forces two somewhat incompatible plant types together and expects optimum performance from each.
2. Trees and turf share the same basic requirements: sunlight, air, water, and minerals. When growing together, they compete,

above and below ground. Besides vying for space, sunlight, and water, each has ways of inhibiting the growth of the other. This phenomenon, known as allelopathy, is difficult to measure.

3. Managing trees in close proximity to manicured turf can present some challenges. Shade on the turf must be minimized without sacrificing the functionality of the trees. Root interference with turf needs to be avoided or eliminated. The turf must be protected when performing tree maintenance.

4. Turf management practices can also adversely affect the health of the nearby trees. Mower and weed trimmer damage is a common problem on golf courses. Excess fertilization and irrigation can cause stress problems for trees. Broadleaf herbicides, which do not harm grasses, can be fatal to trees.

5. Despite the differences between trees and turf, it is possible to achieve a balance in managing both on a golf course. Good planning and selection is critical, reducing potential conflicts before they occur.

6. Management practices such as proper pruning, mulching, and plant health care can minimize competition, reduce injuries, and maximize the potential of both the trees and the turf.

4

Design and Construction

Designing with Trees

When golf course architects set out to design a course, the land upon which the course is to be constructed may or may not have trees. Some courses are formed from land that is almost completely devoid of trees. Other courses are literally carved out of forests. Each case presents a set of opportunities and problems.

If virtually every tree on the course is to be planted, the designers have the benefit of being able to select the exact combinations of trees for every function. In addition, there will not be the concern of preserving trees during construction. Unfortunately, designers often do not take into account the effects the trees will have when they mature. The initial concern is getting the young trees to grow as fast as possible. Too often the end result is tree problems caused by poor planning.

Many courses today are being built on land that is dense with trees. The barriers around the course perimeter and between fairways are already in place. Doglegs can be created immediately and greens have a built-in backdrop. This scenario, however, has its drawbacks. The loss of trees from construction damage over the first five to ten years after construction is usually heavy. Trees that may have been critical in the design are often lost. Without an arborist's consultation, the trees selected to remain may not be the best choices, leading to other problems down the line.

Start with a Plan

Regardless of whether there are trees on the land, the design plans should include forethought about the function of trees on the completed course. If there are existing trees, the first step is to survey the property and map out the location, species, size, and condition of each tree. From this information a trained arborist can work with the golf course architect and builder to decide which trees are to be preserved, which can be transplanted, and which should be removed.

4.1. Trees must be included in the design process. Existing trees that are to be preserved must be protected. New trees must be selected to serve their intended function, and to thrive in their new environment.

In addition to a tree inventory, the property survey must include a site assessment. The site assessment maps out prevailing winds, sun direction, existing topography and plants, and sometimes even a soil survey. Each of these factors must be considered when planning the course because they will influence the choice of species in each site.

The challenge is to combine the functional and aesthetic qualities of what is already present with what is proposed for new construction. The trees play an important role not only in the play of the game, but in the overall feeling of the course. Trees define space in a three-dimensional sense, creating an *implied* enclosure. Although a wide-open course may be ideal for the beginning golfer, most American players prefer the functional and aesthetic qualities of fairways separated by tall trees. Lack of vertical definition can be disconcerting.

When designing with trees it is important to consider the function of each tree or group of trees on the course. For example, if trees are to be used as a windbreak you must consider the height of the trees, the distance from the fairway or green, the species (evergreens are effective year-round), and the density of the planting. Each of these characteristics, however, impacts other aspects such as shade effects on the turf.

4.2. Trees provide the third dimension on a golf course, creating a feeling of enclosure.

The same trees that work so well as a windbreak may block the crucial morning sun that is needed to maintain a thick, healthy green.

When selecting trees consider the potential size at maturity. Take into account the height, spread, shape, and form. Each of these has an effect on the shade that will be cast. Trees should not be planted too close to greens. Maintain a distance of 50–60 feet on the south and east sides of the green so as not to limit the morning sun.

Be sure to plan for species diversity, yet avoid too much diversity in small areas. Repetition of species is both more natural and more pleasing to the eye. Keep in mind that diversity in ages of trees is important as the course matures.

Avoid planting too close to fairway "flight paths." Remember that trees are penal. Consider whether the trees will overhang fairways or tees. Do not block views, but rather, enhance them. Blocking the view of a bunker has the effect of creating a double hazard. Blocking the view of a well-designed hole is self-defeating.

When planning for fairway plantings, resist the urge to fill every gap with trees. Trees should be planted in groups. Small, individual trees look awkward, are harder to maintain, and tend to suffer more stress-related problems. Avoid designing straight lines of trees. Uneven num-

4.3. If possible, keep cart paths at least 6–8 feet from mature trees.

bers of trees in the clusters and uneven spacing of the groups creates a more naturalized look. Plan for sufficient space for maintenance equipment to get through and around the plantings.

Avoid using shrubs for yardage markers. They appear out of scale with the rest of the course, they can be difficult to maintain within the turf, and they tend to create an unreasonable hazard for golfers to negotiate.

The design plan must include a long-range plan for planting and maintaining trees. Most superintendents have come to realize that: (1) trees change the play of each hole as they grow, (2) trees require continued care, and (3) the loss of a significant tree or group of trees can actually mean modifying or redesigning the hole.

Even established courses must have a long-range plan for tree planting and maintenance. It is not unusual for a greens committee to allocate generous funds for tree planting. Without a plan, staff will place the trees wherever space seems available. If the trees survive, they may develop health problems due to poor tree/site matches. They may also create problems for the play of the game as they outgrow their sites. The most negative aspect may be emotional attachments. It is often difficult to obtain authorization to remove a tree, even if it should never have

4.4. Tree planting must be part of the overall management plan for a golf course, taking into account both site considerations and mature tree characteristics.

been planted in the first place. Some superintendents keep a list of wanted plants for members to donate as memorial plantings. This assures that new trees are planted according to the master plan.

Tree Selection

All trees on a golf course serve a function. Some trees delineate the fairways, others create doglegs. Trees form a backdrop behind the greens, helping golfers with their sense of depth perception. Some trees may serve no other function than to shade the players as they wait to tee off. Many of the trees may have existed before the golf course was built. Others are planted as part of the master plan, designed with a purpose in mind.

Golf course architects may consider the selection of trees available in any given area as a palette from which to choose when creating desired effects. If a visual screen is needed, select a species with a dense canopy, perhaps an evergreen. If the tree is to be planted near a green, avoid species that drop fruit, seeds, or twigs. Trees used to form a dogleg must be tall and thick enough so that golfers will not be able to cut the corner.

When deciding what trees to plant in each site, you must consider the desired function of the tree, the cultural requirements and characteristics of each species, and the environmental conditions of the specific site.

Matching Tree and Site

The most important part of tree selection is matching the tree with the site. Every tree species has certain cultural requirements including light, water, soil conditions, growing space, and others. Each planting site has environmental characteristics such as temperature extremes, soil pH, and light levels that can limit which plants will thrive.

With such a variety of site and plant considerations, it is necessary to set priorities in selecting a tree. The highest priorities are those that affect tree survival, such as light and moisture needs. Sometimes a compromise must be made between the functional goals and the site limitations. For example, if a screen is desired, an evergreen species might be selected because it would provide year-round screening. However, if shade is a limiting factor, a deciduous species should probably be selected since only a few evergreens grow well in the shade.

Computer plant selection software programs can make the plant selection process a little easier. These allow the user to generate a list of trees that match the site and the plant characteristics that are desired. For example, the computer could generate a list of trees that are native to a region, drought tolerant, less than 30 feet tall, and have showy, white flowers.

Site Considerations

Many landscape architects do a site analysis before a landscape plan is designed. This analysis records existing site conditions that may affect plant selection. In addition, the functional goals of the design are outlined. This helps the designer select plants that are appropriate. There are many site characteristics that must be taken into consideration in plant selection. Plants may be limited by growing space, water availability and drainage, soil pH, light levels, and weather.

The amount of growing space available is important when selecting the type of tree to plant in a particular site. The area above and below ground must be large enough to allow the tree to reach its mature height, branch spread, and trunk diameter without interfering with surrounding objects. It is easy to forget that many trees can grow over 100 feet in height if conditions are favorable. Many trees' lateral spread exceeds their height. This must be considered to avoid encroachment over fairways or becoming large enough to cast too much shade on greens or tee boxes. The root system cannot be overlooked either. Many superintendents constantly fight surface roots interfering with turf.

The climatic or environmental conditions of the general area must also be considered. This includes not only the climate of the geographical zone, but also the microclimate of the planting site which can be affected by buildings, raised planters, and other surroundings. Light levels may significantly affect plant growth and survival. Shade is a problem to some trees. Many trees cannot survive in a heavily shaded site. Trees that are intolerant of shade are sometimes the best choices for planting in close proximity to turf. These trees tend to have open canopies and allow more sunlight to pass through their crowns to the turf below.

The soil conditions of the site must be examined. It is advisable to get a soil analysis to determine the texture, pH, soluble salts, and nutrient levels of the soil. One of the most important aspects of the soil is its water-holding capacity and drainage. Trees that are planted in a site that is too wet or too dry often die within the first year. If the soil is compacted, tree growth may be greatly reduced due to insufficient oxygen in the root zone. Be alert to soil contaminants such as herbicides that could poison the planting site.

Tree Considerations

There are many factors that must be considered when selecting a tree for a given site. These include growth rate, size at maturity, form, hardiness, insect and disease resistance, and maintenance requirements.

Fast-growing trees are often utilized to provide certain functional uses such as shade and screening. However, rapidly-growing trees often create a heavy maintenance requirement; e.g., to prevent them from encroaching on a green. Trees such as these may lack the amenity value of other species and, in some cases, are more prone to storm damage.

Selecting a tree with a particular growth form can sometimes offer a solution to a growing space problem. For example, a large tree with an upright growth form and narrow branching habit could be planted closer to a building than one with a spreading form. Cultivars of some tree

4.5. Mature trees add to the "atmosphere" of the course, providing shade and beauty.

species are available with different growth forms such as upright, pyra-midal, or weeping. Trees with diverse growth forms provide a variety of architectural effects on the golf course.

Specific plant characteristics may make a tree more desirable. These might include exfoliating bark, attractiveness to birds, or an interesting branching habit. Some trees are admired for their flowers or fall color. Other characteristics, such as excessive leaf, fruit, or twig drop may preclude the use of certain trees adjacent to a green, patio, or parking lot. You should also consider future tree and site management when selecting tree species. Some plants require more water than others, and

**4.6. Species selection is a matter of matching tree and site.
Fast-growing species are often selected to reach a large
size quickly. These trees are sometimes weak-wooded
and prone to storm damage as they mature.**

supplemental irrigation or drainage may be necessary. Certain trees need to be pruned regularly. Others may be prone to storm damage.

With arborists taking a more holistic approach to tree health, more consideration is given to tree selection. Of particular concern is resis-

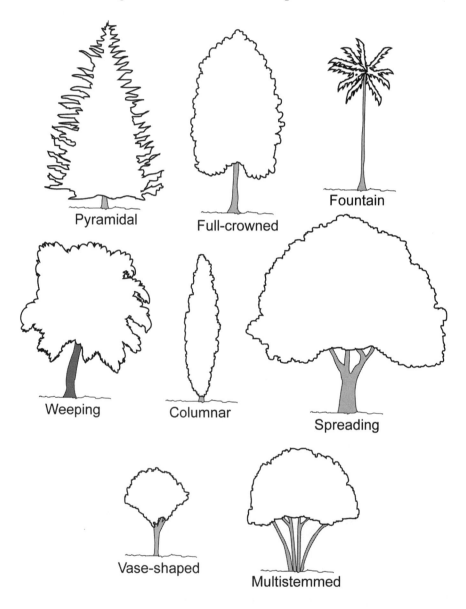

4.7. Various growth forms of trees.

4.8. Some flowering ornamental trees offer a lovely display
before leaf set in the spring.

4.9. Consider late season attributes such as the fruit
remaining on this crab apple.

tance or susceptibility to insects and diseases. It is important to select a tree that will not be stressed on a given site. Stress can weaken a tree's ability to withstand the invasion of certain insects and diseases. Many health problems that affect trees on a golf course can be avoided by better planning and matching the tree to the site.

Avoiding Tree Damage During Construction

The primary focus of this section deals with construction of new courses. The problems, however, are not limited to new course construction. Construction often takes place on existing courses; adding

4.10. Fall color, especially in the northeastern United States, can be a major attraction for golfers in the autumn.

buildings, adding or rerouting irrigation, changing tee boxes, and changing the grade of greens and bunkers.

The processes involved with construction on a golf course can be devastating to the surrounding trees if measures are not taken to protect them. The visible injuries such as broken branches and wounds to tree trunks are only the beginning. It is the damage to the root systems that often result in tree loss. Furthermore, unless the damage is extensive, the trees may not die immediately, but could decline over several years. With this delay in the manifestation of symptoms, you may not associate the loss of the tree with the construction.

It is possible to preserve trees during golf course construction if proper measures are taken. This is one area where it would be advisable to hire a consulting arborist for planning advice if you do not have an arborist on staff. An arborist can help you decide which trees can be saved and can work with the construction contractors to protect the trees throughout each construction phase.

4.11. Tree selection includes the specimens as well as the species. These trees are sugar maples. The one in the foreground has poor branching structure, probably resulting from poor maintenance practices in the nursery.

Damage Caused by Construction

Physical Injury to the Trunk and Crown

Construction equipment can injure the above ground portion of a tree by breaking branches, tearing the bark, and wounding the trunk. These injuries are permanent, and if extensive, can be fatal.

Cutting of Roots

The digging and trenching necessary to construct a clubhouse, install an irrigation system, or install underground utilities will likely sever a portion of the roots of many trees in the area. It is easy to appreciate the potential for damage if you understand where roots grow. The roots of a mature tree extend far from the trunk of the tree. In fact, roots typically will be found growing a distance of 1–3 times the height of the tree. The amount of damage a tree can suffer from root loss depends, in part, upon how close to the tree the cut is made. Severing one major root can cause the loss of 5 to 20 percent of the root system.

Another problem that may result from root loss due to digging and trenching is that the potential for the trees to fall over is increased. Roots anchor a tree in the soil. If the major support roots are cut, the tree may fall or blow over.

Soil Compaction

An ideal soil for root growth and development is about 50 percent pore space. These pores, the spaces between soil particles, are filled with water and air. The heavy equipment used in construction compacts the soil, and can dramatically reduce the amount of pore space. This not only inhibits root growth and penetration, but also decreases oxygen in the soil that is essential to the growth and function of the roots.

Smothering Roots by Adding Soil

Most people are surprised to learn that 90 percent of the fine roots that absorb water and minerals are in the upper 6–12 inches of soil. Roots require space, air, and water. Roots will grow best where these

4.12. Before installing a new cart path, it is important to consider the impact on the root systems of nearby trees. Damage such as this can kill some trees, or render them susceptible to blowing over in a strong wind.

4.13. Trees are intolerant of construction fill to raise the grade. The species and maturity of the tree, as well as the depth of the fill, will determine how fast the tree will die.

requirements are met, which is usually very near the soil surface. Piling soil over the root system or increasing the grade will smother the roots. It only takes a few inches of added soil to kill a sensitive, mature tree.

Exposure to the Elements

Trees in a forest situation grow as a community, protecting each other from the elements. The trees grow tall, with long, straight trunks and high canopies. Removal of neighboring trees or opening the shared canopies of trees will expose the remaining trees to sunlight and wind. The higher levels of sunlight may cause sunscald on the trunks and branches. Also, the remaining trees will be more prone to damage from wind or ice loading.

Compounded Damage

Unfortunately, trees are likely to suffer from multiple forms of damage as a result of construction. Even if measures are taken to protect the trees, the root systems will probably suffer from compacted soil and grade changes.

Preserving the Trees

One of the first decisions is determining which trees are to be preserved and which should be removed. You must consider the species, size and maturity, location, and the condition of each tree. The largest, most mature trees are not always the best choices to conserve. Younger, more vigorous trees can usually survive and adapt to the stresses of construction better. Try to maintain diversity of species and tree maturity. Your arborist can advise you about which trees are more sensitive to compaction, grade changes, and root damage.

Planning

Your arborist and contractor should work together in planning the construction. The contractor and all subcontractors must be educated regarding the value of the trees on the golf course and the importance of saving them. Few are aware of how and where tree roots grow, and what is needed to preserve them.

Sometimes small changes in the design of a fairway can make a great difference in whether a critical tree will survive. An alternative plan may be more friendly to the root system. For example, bridging over the roots of a crucial tree instead of digging down to install a cart path may save the tree. Instead of trenching beside a tree for utility or irrigation installation, tunneling under the root system. It is less damaging.

4.14. Digging, trenching, and grade changes within the root system of a tree can lead to tree decline or death. In addition, heavy equipment compacts the soil, causing further stress to the roots.

Erecting Barriers

Because you'll have limited options for treating and repairing construction damage later, it is vital that the trees be protected. The single, most important action that you can take is to set up construction fences around the trees that are to remain. The fences should be placed as far out from the trunks of the trees as practical. The intent is not merely to protect the above ground portions of the trees, but also the root systems. Remember that the root systems extend much farther than the driplines of the trees.

Instruct the construction personnel to keep the fenced area clear of building materials, waste, and excess soil. No digging, trenching, or other soil disturbances should be allowed in the fenced area.

Limited Access

If at all possible it is best to allow only one access route on and off the property. All contractors must be instructed where they are permitted

4.15. If trees are to be preserved during golf course construction, they must be protected. Erecting a fence around the root system is one facet of tree preservation.

to drive and park their vehicles. Often this same access drive will later serve as the route for utility wires, water lines, and/or the cart path.

Limit storage areas for equipment, soil, and construction materials. Specify areas for burning (if permitted), wash-out pits, and construction work zones. These areas should be away from protected trees.

Specifications

Get it in writing. All of the measures intended to protect your trees must be written into the construction specifications. The written specifications should detail exactly what can and cannot be done to and around the trees. Each subcontractor has to be made aware of the barriers, limitations, and specified work zones. It is a good idea to post signs as a reminder.

Fines and penalties for violations should be built into the specifications. Not too surprisingly, subcontractors are much more likely to adhere to the tree preservation clauses if their profit is at stake. The severity of the fines should be proportional to the potential damage to the trees and should increase for multiple infractions.

Maintaining Good Communications

It is important to work together as a team. You may share clear objectives with your arborist and your primary contractor, but one subcontractor can destroy your prudent efforts. Construction damage to trees is often irreversible.

Visit the site at least once a day if possible. Your vigilance will pay off as workers learn to take your wishes seriously. Take photos and videos at every stage of construction. If any infraction of the specifications does occur, it will be important to prove culpability.

Final Stages

It is not unusual to go to great lengths to preserve trees during construction, only to have them injured during the final stages. Installing irrigation systems and rototilling planting beds are two ways the root systems of trees can be damaged. Remember that small increases in grade can be devastating to your trees. Careful planning and communicating with contractors is just as important as avoiding tree damage during construction.

Treatment of Trees Damaged During Construction

There are some remedial treatments that may save some construction-damaged trees, but immediate implementation is critical. If you have trees that have been affected by recent construction, consult a professional arborist promptly. Your arborist can assess each tree for viability and potential hazards, and recommend treatments.

Inspection and Assessment

Since construction damage can affect the structure and stability of a tree, your arborist should check for potential hazards. This may involve a simple visual inspection, or instruments may be used to check for the presence of decay. Sometimes the hazard can be reduced or eliminated by removing an unsafe limb, pruning to reduce weight, or installing cables or braces to provide structural support. An often-overlooked method of reducing hazards is to move objects that could be hit, or to limit access to the hazardous area. If there is doubt about the structural integrity of a tree, or the hazard cannot be adequately reduced, it should be removed. Although the goal is to preserve the trees whenever possible, that goal must not supersede any question of safety.

Treating Trunk and Crown Injuries

Pruning

Branches that are split, torn, or broken should be removed. Also remove any dead, diseased, or rubbing limbs from the crowns of the trees. Sometimes it is necessary to remove some lower limbs to raise the canopy of a tree and provide clearance below. It is best to postpone other maintenance pruning for a few years.

Old recommendations suggest that the tree canopies should be thinned or topped to compensate for root loss. There is no conclusive research to support this practice. Thinning the crown can reduce the tree's food-making capability by removing foliage and may stress the tree further.

It is better to limit pruning in the first few years to hazard reduction and the removal of deadwood. **Do not top the trees.**

Cabling and Bracing

Trees growing in wooded areas are usually not a threat to people or structures. Trees that are close to clubhouses or other buildings must be maintained to keep them structurally sound. If branches or tree trunks need additional support, a professional arborist may be able to install cables or bracing rods. If cables or braces are installed however, they must be inspected regularly. The amount of added security offered by the installation of support hardware is limited. Not all weak limbs are candidates for these measures.

Repairing Damaged Bark and Trunk Wounds

Often the bark may be damaged along the trunk or major limbs. If this happens, remove the loose bark. Jagged edges can be cut away with a sharp knife. Take care not to cut into living tissues.

Wound Dressings

Wound dressings were once thought to accelerate wound closure, protect against insects and diseases, and reduce decay. However, research has shown that dressings generally do not reduce decay or speed closure, and rarely prevent insect or disease infestations. Most experts recommend that wound dressings not be used. If a dressing must be used for cosmetic purposes, use just a thin coating of a nontoxic material.

Irrigation and Drainage

One of the most important tree maintenance procedures following construction damage is to maintain an adequate, but not excessive supply of water to the root zone. If there is a drainage problem, the trees will decline rapidly. This must be corrected if the trees are to be saved. If soil drainage is good, be sure to keep the trees well watered especially during the dry, summer months. A long, slow soak over the entire root zone is the preferred method of watering. Keep the top 12 inches moist, but avoid overwatering. Avoid frequent, shallow waterings. Make sure surface water drains away from the tree. Proper irrigation may do more to help the trees recover from construction stress than anything else you could do.

Mulching

One of the simplest and least expensive things you can do for your trees may also be one of the most effective. Applying a three to four inch layer of an organic mulch such as wood chips, shredded bark, or pine needles over the root system of a tree can enhance root growth. The mulch helps condition the soil, moderates soil temperatures, maintains moisture, and reduces competition from weeds and grass. The mulch should extend as far out from the tree as practical for the landscape site. (If the tree had a say, the entire root system would be mulched.) Do not apply the mulch any deeper than four inches, and do not pile it against the trunk.

Improving Aerating of the Root Zone

Drilling Holes/Vertical Mulching

Compaction of the soil and increases in grade have the effect of depleting the oxygen supply to tree roots. If soil aeration can be improved, root growth and water uptake can be enhanced. The most common method of aeration of the root zone involves drilling holes in the ground. Holes are usually 2–4 inches in diameter and are made about 3 feet on center, throughout the root zone of the tree. The depth should be at least 12 inches, but may need to be deeper if the soil grade has been raised. Sometimes the holes are filled with peat moss, wood chips, pea gravel, or other materials that maintain aeration and support root growth. This is called vertical mulching. Although helpful in some situations, vertical mulching is limited in the extent that it can improve aeration of the root zone.

Radial Aeration

More recent research has shown promising results with another method of aeration called radial aeration. Narrow trenches are dug in a radial pattern throughout the root zone. These trenches appear similar to the spokes of a wagon wheel. It is important to begin the trenches 4–8 feet from the trunk of the tree to reduce the chances of cutting any major support roots. The trenches should extend at least as far as the dripline of the tree. If the primary goal is to reduce compaction, the trenches should be about 1 foot in depth. They may need to be deeper if the soil grade has been raised. This technique is appropriate for isolated trees, where the roots of surrounding trees would not be damaged.

The narrow trenches can be backfilled with the improved natural soil or compost. Root growth will be greater in the trenched area than in the surrounding soil. This can give the tree the added boost it needs to adapt to the compacted soil or new grade.

Vertical mulching and radial trenching are techniques that may improve conditions for root growth. If construction-damaged trees are to survive the injuries and stresses they have suffered, they must replace the roots that have been lost.

What about Fertilization?

Most experts recommend that you do not fertilize your trees the first year after construction damage. Water and mineral uptake may be reduced due to root damage. Excessive salts from the fertilizer can draw water out of the roots and into the soil. In addition, nitrogen fertilization may stimulate top growth at the expense of root growth. It is a common misconception that applying fertilizer gives a stressed tree a much-needed shot in the arm. Fertilization should be based on the nutritional needs of the trees in a site. Soils can be analyzed to determine whether any of the essential minerals are deficient. If soil nutrients are deficient, supplemental fertilization may be indicated. It is advisable to keep application rates low, however, until the root system has had time to adjust.

Monitoring for Decline and Hazards

Despite your best efforts you may lose some trees from construction damage. Symptoms of decline include smaller and fewer leaves, dieback in the crown of the tree, and premature fall color. If a tree dies as a result of root damage it may be an immediate hazard and should be removed right away. Examine your trees for signs of possible hazards. Look for cracks in the trunk, split or broken branches, and dead limbs. Watch for indications of internal decay such as cavities, carpenter ants, soft wood, and mushroom-like structures growing on the trunk, root crown, or along the major roots. If you detect any defects or suspect decay, consult an arborist for a professional assessment. It is prudent to have your trees evaluated periodically by a professional.

You should also inspect your trees for signs of insects or diseases. Stressed trees are more prone to attack by certain pests. Talk to your arborist about putting your trees on a program of Plant Health Care (PHC). This may help identify and treat problems before they become a threat to the life of your trees.

When to Consider Transplanting Trees

A common problem that most golf course designers face is that specimen trees may be growing in locations where they simply cannot stay. An uncommon option is to leave the tree anyway. There are a few cases of trees actually growing on a fairway or green. Although this certainly creates an unusual feature, it can be a nightmare for golfers and superintendents alike. A practical alternative to removing trees may be transplanting.

There are several factors that must be considered when deciding whether to transplant a tree. From a survivability point of view you must take into account the size, species, condition, and location of the tree, as well as the time of year. Other, more practical aspects include the value of the tree, cost of transplanting, location and distance, and potential damage to the course from moving equipment.

Some species transplant more easily than others do. Most species of palms are easier to move than deciduous trees or conifers. The ease of transplant has an obvious impact on the size of tree that can be moved. Tree spades that can dig and lift very large trees are available. Transplanting large trees requires much experience and proper equipment. If you have one or two years of advance notice, you can increase the chances of tree survival by root pruning. Also, trees generally have a much higher survival rate if moved while they are dormant.

When deciding whether to attempt moving a tree, the first determination is the survivability. If tree survival is in doubt, a decision must be mutually understood and agreed upon. It is difficult to guarantee survival. In addition, the tree must have enough value in the new location, either due to function or aesthetic considerations to justify the expense of moving it. The expense will depend on the size of the tree and the distance of the move. Unless the tree is fairly small, this is one operation that is best left to specialists.

A final, very important component to the decision is the potential damage to the course. This may not be a significant factor if the course is under construction. If the course is established, it may be the limiting factor. Large tree transplanting requires large equipment. This equipment can injure adjacent trees, destroy turf, and damage underground irrigation systems. In temperate climates trees are sometimes moved after the ground is frozen, reducing problems of turf damage. No matter what location is involved, take the time to have utilities and irrigation systems marked before proceeding.

Summary

1. Trees must be considered when designing new golf courses, constructing courses, and when modifying existing courses.

4.16. Although there are tree spades that can move relatively large trees, the decision whether to attempt transplanting must consider all factors. Forest-grown trees can be especially difficult to transplant.

Each scenario has its advantages and disadvantages. If new trees are being planted, planners have the ability to select the most appropriate species for the site and function. Too often, though, designers do not take into account the effects the trees will have when they mature.

2. Some courses are literally carved out of woodlands. Although the backdrops, doglegs, and course perimeters are virtually immediate, tree loss in the first few years after construction can be heavy. With the help of a good consulting arborist, these losses can be minimized.

3. It is important to survey and inventory the site. The number, location, size, species, and condition of each tree should be recorded. Also, map out the site conditions such as sun direction, prevailing winds, existing topography, etc. before beginning planning or construction.

4. When designing with trees consider the function of each tree or group of trees on the course. Consider the potential size of each tree at maturity. Select trees that will function well in the desired site. Establish a long-range plan for sustainability.

5. The processes involved with construction on a golf course can be devastating to the nearby trees if measures are not taken to protect them. Visible injuries such as broken branches and wounds to tree trunks are only the beginning. It is the damage to the root systems that often results in tree loss. A consulting arborist can help decide which trees can be saved and can work with planners and construction contractors to protect the trees during construction.

6. Unfortunately, the treatments for trees damaged by construction are very limited. Although there are a few remedial measures that may help the stressed trees, it is far better to invest in preserving the trees by avoiding damage during the construction phase.

5

Planting and Transplanting

Golf course superintendents tend to plant a large number of trees. It can be very difficult, however, to get young trees established on courses. Mechanical damage can be a major problem. If you can keep the mowers away, you're ahead. Bark injury and bruising from golf balls is another matter. Proper watering may be the biggest problem in getting newly planted trees established. Don't assume that the irrigation system will keep the trees adequately watered. After planting, the available water within the root ball is depleted rapidly. Irrigation to surrounding turf is unlikely to fulfill the need of the roots within the ball, and if the tree is on the downhill part of the grade, or if irrigation rates are high, water may accumulate in the planting pit and drown the tree.

The key to giving a tree a healthy start is good planting procedures. Stress and physiological disorders can often be traced to poor planting practices. Many tree-planting traditions have been passed down through generations in the landscape and arboriculture industry. While some of these techniques are still recommended today, others have been changed to reflect current research and technology. A well-informed golf course superintendent should be aware of the latest techniques in tree planting and transplanting.

Selecting Quality Stock

How to Select a Healthy Plant

For a tree to be a success when transplanted onto a golf course, it is important to begin with a healthy plant. Nursery-grown trees are usually superior to trees collected from other sites; e.g., woodlands, since more of the roots are contained in the root ball.

Look for trees exhibiting good twig extension growth. Trees that have good branch spacing and trunk taper are more desirable than those that have been headed back. Foliage should be evenly distributed on the upper two-thirds of the tree and not concentrated at the top. Avoid trees with many upright branches, or with twin or codominant leaders. Healthy plants establish quickly in the landscape. Plants in poor health attract pests and require more maintenance. Examine the leaves and shoots. Choose trees with an abundance of healthy, green leaves without signs of chlorosis. Check for the presence of insects or disease.

Except for small-growing, multistemmed ornamentals, select trees with a single trunk and leader, and spreading branches. Check the plant for mechanical damage. Do not purchase a tree that has an injury to the trunk or numerous broken branches. Examine the root ball of the tree. Balled and burlapped (B&B) plants should have a solid ball that has been kept moist and protected from drying. If the plant is in a container, check the root system. Roots that are brown or black indicate a health problem. Avoid trees with circling roots that may develop into girdling roots.

A point that cannot be overemphasized is the importance of matching the tree and its requirements to the planting site conditions. The best planting procedures known will not save a tree that is poorly suited for its site. The tree must be able to tolerate site conditions such as wet or dry soils, soil pH, growing space, and full sun or shade. Selecting a tree that meets the site requirements is the single most important factor in influencing the success of the plant.

Develop a Relationship with a Nursery

A wise golf course superintendent will develop a strong relationship with a good production or wholesale nursery. Choose a nursery that provides a diverse species selection with a moderate to large volume of stock. Get to know the owner or manager. This relationship will allow more control of the final product. If you are buying in large quantity, you may be able to specify maintenance, training, and digging requirements for the stock you need. This will enable you to select trees that are well suited for a golf course planting. You will probably be permitted to handpick (tag) stock from the field, and you should qualify for quantity discounts.

You can plan for your long-term needs with your nursery. You may be able to request that some hard-to-find species be grown if you can guarantee their purchase. Your nursery connections can keep you informed of the latest selections, cultivars, and varieties. They may even be willing to grow your course "signature" trees.

Of course there is always the option of maintaining a small nursery on your own grounds. This may allow you to supplement your nursery stock with some speciality species. Although you have total control over species and growing specifications, you must have the trained personnel to maintain the stock. Another potential disadvantage is a weakened relationship with your nursery. They may be less likely to offer you the same privileges if you grow much of your own stock.

Site and Hole Preparation

Once the tree has been selected, the planting pit (hole) can be dug. The planting pit must allow for rapid root development without restriction. The planting pit should be much wider in diameter than the root ball. If the soil is compacted, the hole should be at least two to three times the width of the root ball. Most of the root growth will be shallow and horizontal. A wide planting pit with sloped sides will allow the roots to spread.

Never dig the hole deeper than the root ball, and never put soft backfill in the bottom of the hole before planting. The bottom of the hole must be firm. The planting pit may act as a dish and hold water, especially in clay soils. Oxygen levels are low in the bottom of such holes, and not conducive to rapid root growth. The tree must be planted at the right depth with the root collar at the soil surface. Planting a tree too deep can stress the tree and drown or exclude necessary air to the roots. If backfill is put into the bottom of the planting pit, it must be tamped down firmly to prevent the root ball from settling too deep.

An often-overlooked method of digging the planting pit is to use a stump grinder. This machine can create a wide, sloping pit while keeping the bottom firm. In addition, the natural soil is pulverized, creating an ideal backfill. Many tree service companies will contract to do this work.

Adequate drainage is an important consideration in successful planting. Poor drainage, characteristic of heavy, clay soils, accounts for many transplant losses. Since there has often been a great deal of land sculpting on a golf course site, there may be areas that have been stripped to the subsoil and others that have extra fill. It is a good idea to check drainage and water percolation rates before planting. When planting

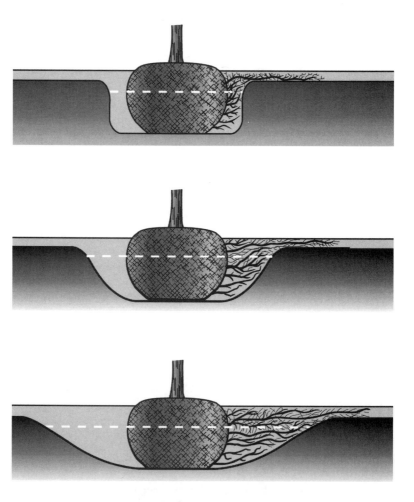

5.1. A wide planting pit allows better development of new roots than a traditional hole or a planting hole with sloped sides.

large trees in poorly drained soils, install a pipe drainage system. Adding gravel in the bottom of the hole creates a "perched" water table as water accumulates in the soil above the gravel. This actually makes the drainage problem worse.

How to Plant a Tree

The root ball of the tree must be handled carefully. Small, fibrous absorbing roots are easily broken, and it is advisable to check the handling of plants during delivery. Place the tree in the planting hole gently. Check to see that the root collar will be no deeper than the soil level. Sometimes trees will have been planted too deep in the nursery. Be sure to locate the root collar and remove soil from the top of the ball, if necessary. Planting too deep is the most common mistake and can lead to a slow death of the tree. Where drainage is poor, it may be advisable to plant the tree 1–4 inches higher than natural grade. If practical, the tree should be oriented in the hole so that it faces the same direction as it did before it was dug. In certain climates, thin-barked species, such as maple and beech, are susceptible to sunscald.

In most cases it is acceptable to backfill the pit with the same soil that came out of the hole. Research has shown that soil amendments do not necessarily assist in tree establishment and growth. If the existing soil is extremely poor, amending or replacing the soil may be best. However, the backfill soil type should match the soil type of the site as closely as possible. Backfilling with a sandy loam in heavy clay may cause the planting hole to collect water and suffocate the roots. If the backfill must be amended, a well-decomposed compost or other organic matter may help improve soil structure and fertility. Add about 25 percent by volume, and try to amend as large an area as possible to provide textural continuity. Where soil textures differ, natural water movement may be impeded.

Work the soil around the ball so that no air pockets are left. The soil in the lower quarter of the hole, around the base of the root ball, should be tamped firmly to help support the tree. Large pockets of air can allow the roots to dry out. Firm the soil so that the tree is vertical and adequately supported, but do not pack the soil. Water thoroughly, in stages, while backfilling.

Sometimes it may be helpful to create a saucer around the tree to collect water over the root zone, especially on sloped sites. This will

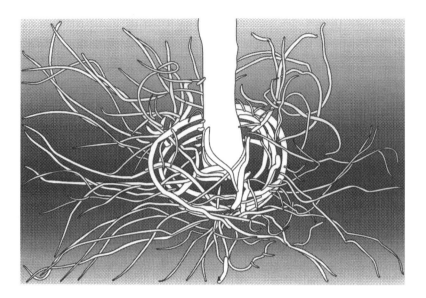

5.2. Circling roots that can develop inside containers
may become girdling roots later.

5.3. Girdling roots are often the result of circling roots in
containerized nursery stock or poor planting practices.

allow water to percolate into the root ball where it is needed. With heavy, clay soils, excess water may accumulate during wet periods, however, and the saucer may have to be opened to allow water to drain away.

Natural burlap is biodegradable, but some nurseries use treated burlap or synthetic wraps. It is good practice, whatever the material, to remove the wrap from the top and sides of the root ball. Often the wrap is held in place with twine. All twine or ropes tied around any part of the tree must be removed to avoid girdling the tree. Remove all tags or labels so that they will not girdle the trunk or branches as the tree grows.

Some larger balled trees come in wire baskets to maintain the integrity of the ball in handling. If it is impractical to remove the entire basket, it is preferable to cut away as much of the wire as possible, once the tree is in the planting pit. This eliminates interference with rakes or lawn mowers if the tree is planted shallow and allows roots to grow and spread freely near the surface without being impeded by the wires.

Transplanting

Transplanting a tree involves the additional procedures of digging and preparation for moving. Some species transplant more easily than

5.4. Newly planted trees should have a raised ring of soil or mulch to direct water down to the root ball. Avoid placing the berm farther out than the radius of the ball.

others. Since digging a tree for transplanting can remove as much as 95 percent of the absorbing roots, trees that are difficult to transplant should be moved when conditions are optimal.

In general, the best time to transplant most tree species is in the early spring or autumn. Many deciduous plants can be moved just after leaf drop in the autumn, while the moisture level in the soil is relatively high. If the soil is warm enough, roots have a chance to grow and begin to establish before the ground freezes. Some plants are more easily transplanted in the spring before bud break. Transplanting dormant trees reduces demand for soil moisture since transpiration is minimal. Evergreen trees are also more easily transplanted while dormant.

If you know that you will be transplanting a tree in the future, you have a big advantage. You can increase the percentage of roots in the root ball by root pruning a year or two in advance. Prune the roots back to one root ball size smaller than the size that will be moved. The remaining roots will generate prolific smaller roots from the cuts. Then, when you dig a larger ball for transplanting, the root system will have a head start and the tree will recover faster. Of course, the extra roots on the root ball will only be a benefit if the water supply is adequate after planting.

The techniques used in digging vary little with species, size, and soil type. The procedures described here are for digging and transplanting a large tree. The techniques are similar for smaller trees that are easier to move. Before digging, tie the lower branches of the tree to prevent injury or breaking. Care must be taken to avoid damage to the bark. Do

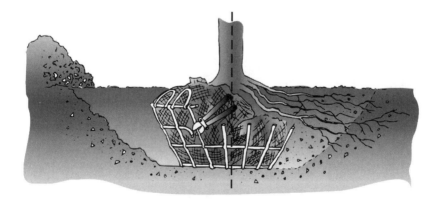

5.5. Remove the upper half of wire baskets to prevent future problems. It is also best to fold back the burlap after removing all ties and twines.

not tie the limbs so tightly that a sharp bend is created that could compress the tissues or break the limbs. Protect the trunk with foam, tree wrap, burlap, or other material while transplanting. Be sure to remove this covering once the tree is planted.

Measure the tree 12 inches above the root crown to determine the trunk diameter. The root ball should be at least 10 inches in diameter per inch of trunk diameter. Thus a 6-inch tree should have a root ball that is 60 inches (5 feet) in diameter. The depth of the root ball may vary with the root growth of the tree. Root growth is deeper in certain species than others, and a drier soil tends to cause deeper rooting. In general, a root ball depth of two-thirds to three-fourths of the root ball diameter is recommended.

Make clean cuts around the perimeter of the root ball to avoid tearing or breaking the roots. If the ball is being dug with machinery, the initial ball should be dug several inches larger than the final ball size. The shaping and final root cuts should then be done by hand. As larger roots are located, cut them with loppers or a hand saw. Cut the roots both at the edge of the ball and the outer edge of the digging trench to clear the trench for more digging. While digging the trench, avoid standing on the root ball. The edge of the ball could break down and damage the roots. Use burlap on the sides and across the top of the ball to keep it from drying. For additional support, secure large soil balls with straps or rope over the burlap.

Once the ball is roped and secured, the tree can be undercut. A clean cut can be made using a cable without having to tilt the ball of soil. The cable is pulled using a winch to undercut the ball. Large trees must be removed from the hole with a crane or large tree mover. Place chains around the ball and attach them to the crane hook. Do not lift trees by the trunk as this can cause trunk injury and serious damage to the root ball. Once out of the hole, fasten the burlap to the bottom of the ball.

If the tree must be transported to a distant site, it should be protected. Pad the trunk to protect it from injury. The crown of the tree may be loosely wrapped with burlap to minimize drying and wind damage. Special permits may be necessary to transport large trees on public roads.

Digging and Planting with a Mechanical Tree Spade

If the tree is being moved from one location on the golf course to another on the same course, it is most economical to use a tree spade.

Tree spades are available in various sizes. No attempt should be made to transplant trees larger than the size limitations of the spade. Keep in mind the recommended ratio of root ball size to tree diameter.

Tree spades are often used to dig planting holes. However, holes dug with tree spades tend to be glazed on the sides. This may inhibit root penetration into the surrounding soil. If tree-spade-dug trees are planted in holes dug with the same spade, there will be little area for new root development within the hole. This practice is not recommended, but if tree spades are used to dig the hole, widen and slope the slides of the hole. It may be necessary to roughen the sides of the hole as well.

If tree spades are used for transplanting, care must be taken to obtain a vertical root ball and to plant the tree upright. Planting on a slope can be a problem since the ground level will be different. Remember that the tree must be planted vertically in the new site.

Root system reestablishment takes longer for large trees than for smaller trees. It is not unusual for trees transplanted at a smaller size to recover and outgrow larger trees transplanted at the same time. In temperate climates, the rule of thumb is that it will take one year to reestablish for each inch of tree diameter. Trees become established much

5.6. Tree spades can be used to transplant relatively large specimens. Hole preparation and good planting procedures are important to increase chances of survival.

more quickly in tropical climates because there is a longer season for root growth.

To Stake or Not to Stake

Staking of newly planted trees is not always necessary. In fact, staking can have detrimental effects on the development of a tree. When compared to trees that have not been staked, staked trees produce less trunk taper, develop a smaller root system, and are more subject to breaking or tipping after stakes are removed. In addition, staked trees may be injured or girdled by the staking materials.

Some trees, however, cannot stand upright without support. In windy sites, in sandy soils, or when trees are tall, staking may be required to hold the plant upright until it can support itself. In other instances, staking may be recommended to reduce movement of the root ball and subsequent damage to the fine, absorbing roots. In addition, stakes are sometimes used to protect young trees from mechanical (equipment) damage.

One, two, or three stakes may be used to support a tree. If a single stake is used, it should be placed on the upwind side of the tree. The guying material used to attach the tree to the stake should be broad, smooth, and somewhat elastic. The tree should be tied with a figure eight loop between the tree and the stake to allow for flexibility. The tie should be attached near the top of the stake. If the tree is tied too rigidly to the stake, the tree will develop a less sturdy root system and may be more subject to girdling and breakage above the tie. If two support stakes are used, a single, flexible tie attached to the top of each stake will be sufficient for support. The stakes should be approximately one-third the height of the clear stem. Triple staking provides more protection against wind, lawnmowers, and vandals, but may limit tree movement too much or cause damage at the attachment points.

Trees greater than four inches in diameter are often supported with guy wires. Trees are generally guyed with three or four wires that are anchored in the ground. Anchoring devices include stakes, land anchors, and deadmen. If stakes are used, they should be driven in line with the guy wire, pointing toward the tree. Stakes that are driven perpendicular to the guy wire tend to loosen.

Guy wires are usually passed through sections of hose to protect the tree. The wires and hose are passed around the tree at crotches, and the wires are twisted to tie them off. The guy wires must not be tied tightly around the tree trunk as this could cause girdling. These hose attachments can still cause damage and should be removed as soon as practical. Do not leave them on more than one year.

On larger trees, guy wires can be attached to the tree with eye screws. This does some damage to a tree but eliminates the chance of girdling by guy wires. The size of the eye screw or eyebolt used should be proportionate to the size of the tree. If the eye screws cannot be removed readily when the guy wires are removed, they should be left in place.

Staking or guying systems should be checked regularly to be sure they are not injuring the trees. Support stakes or guy wires should generally be removed after one growing season. If support systems are left in place for more than two years, the tree's ability to stand alone may be reduced, and the chances of girdling injury are increased.

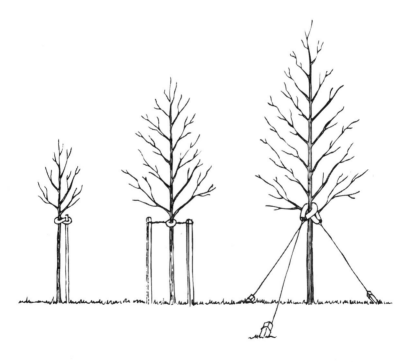

5.7. Three methods of staking and guying trees. Be careful to avoid damage to the branches or trunk. Most cables should be removed after one year.

Maintenance at the Time of Planting

Transplanting is a major operation from which most trees recover slowly. A major portion of the root system is lost in digging, and the tree must reestablish sufficient roots to sustain itself. The tree's ability to obtain and transport water and minerals is greatly reduced. Varying degrees of water stress result, and the tree experiences transplant shock.

Since the root system of a newly planted tree is limited, tree growth will be most limited by drought stress. The tree will live mainly off its energy reserves for at least the first two seasons. Because of this, fertilization is usually not recommended at the time of planting. Excessive fertilizer salts in the root zone can be damaging. If any fertilizer is used, it should be a slow release form.

Pruning following planting should be limited. In fact, a tree will grow and establish most rapidly if pruning is minimized at planting. Dead or

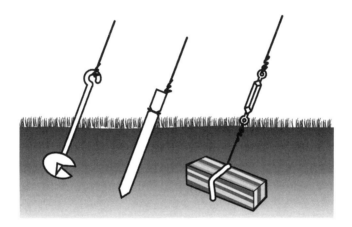

5.8. Three ground anchor systems for guy wires.

5.9. Use a flexible, nongirdling means of attachment.

damaged limbs should be removed. The spacing, balance, and attachment of the limbs should be evaluated. After a two-year establishment period, formative pruning may be desirable. Some limbs may be removed for structural stability or because they are obviously weak or dead. Pruning should be kept to a minimum to retain as much of the leaf area as possible for photosynthesis.

Many early references recommend wrapping the trunks of newly planted trees to protect against temperature extremes, sunscald, boring insects, and drying. More recent research indicates that temperature differentials at the bark are greater with tree wrap than without. Furthermore, tree wrap tends to hold moisture on the bark and can lead to conditions which encourage fungal problems. Also, insects tend to burrow between the bark and the wrap, and can be worse with wrap than without it. Wraps cover the cortex in the bark and can reduce the ability of the tree to photosynthesize. Often young trees on golf courses need protection from mowers and golf balls. Some golf course superintendents use PVC pipe to protect the tender trunks. This is quite effective, but be sure to use perforated pipe or similar tree trunk guards to maintain air circulation. Be sure the guards are loose around the trunk.

5.10. Trees planted onto golf courses are often placed in a stressful environment. Good early maintenance such as water management, mulching, and protection from injury are critical.

The area around the tree should be mulched with 2–4 inches of organic mulch. There is a tendency among maintenance personnel to pile the mulch far too deep. This can lead to health problems for the tree and should be avoided. Properly applied mulch will help reduce competition from weeds and grasses, conserve soil moisture, and moderate soil temperature extremes. The mulch should not be placed against the stem of the tree as that can cause bark suffocation, root crown rot, or rodent damage. The mulch should cover as wide an area as practical since it reduces competition with the turf for moisture and nutrients. Black plastic should not be placed under the mulch since it restricts water movement and oxygen availability to the roots.

Proper watering can be the key to survival of newly planted trees. If rainfall is not sufficient, the tree should be watered. The watering frequency will depend upon temperatures, humidity, soil type, and drainage. Always check the moisture level of the root ball itself, since this may vary from the surrounding soil. A slow gentle soaking of the root zone is the preferable method of applying water. Be sure that the water is getting into the root ball. Excess water accumulation in the planting

5.11. Perhaps the biggest mistake when planting trees is planting too deep. Be sure the trunk flare is above ground. A few inches of mulch is invaluable; a volcano of mulch is not. Be sure water reaches into the root ball when watering newly planted trees.

hole, however, is a leading cause of transplant death. Watering must be appropriate for soil type and drainage.

Summary

1. Golf course superintendents tend to plant a large number of trees. With mechanical damage, irrigation irregularities, and turf maintenance, however, it can be difficult to get young trees established. The key to giving a tree a healthy start is good planting procedures.
2. Start with healthy trees that have a good root ball and strong, well-spaced scaffold branches. Select trees with no mechanical injuries, and avoid damage during transport and planting. Always match the species to the site.
3. Dig a planting pit that is wider, but not deeper, than the root ball. Planting too deep is a common mistake that can lead to a slow death of the tree. In most cases it is best to backfill with the same soil that came out of the hole. Remove all twines, treated or synthetic burlaps, and other materials that can constrict growth.
4. Transplanting a tree involves the additional procedures of digging and preparation for moving. Digging a tree for transplanting can remove as much as 95 percent of the absorbing roots; thus, proper care and timing are important.
5. Root system reestablishment takes longer for large trees than for smaller trees. It is not unusual for trees transplanted at a smaller size to recover and outgrow larger trees transplanted at the same time.
6. Staking of newly planted trees is not always necessary. In fact, staking can have detrimental effects on the development of a new tree. If stakes are required, be sure to use systems that do not girdle the tree. Generally, stakes should be removed after one growing season.
7. Postplanting care is essential. Since the root system of a newly planted tree is limited, water stress is the greatest concern. Make sure the root ball is kept moist, but avoid drowning the tree. Mulch helps reduce moisture stress, moderates soil temperature extremes, reduces competition from weeds and grass, and helps to avoid damage from mowers and trimmers. Pruning and fertilization at planting time should be kept to a minimum.

6

Keeping Trees Healthy

Trees on golf courses present some intense plant health care scenarios. The trees are vital both functionally and aesthetically to the appearance and play of the course, yet their management and care invariably comes second to that of the turf. Unfortunately, the turf care and the tree care can be mutually detrimental. The trees may suffer from mower and trimmer damage, herbicide injury, excess fertilization and irrigation-related problems. Superintendents struggle to maintain the turf while battling shade, surface roots, and debris from nearby trees. Moreover, the tolerance for health problems of either trees or turf is very low on most golf courses.

The challenge for the superintendent is to maintain both turf and trees in good health without one undermining the other. Fortunately, trees and turf share some basic biological requirements (sunlight, water, minerals, and aerated soil). If a balance can be achieved in providing near optimum growing conditions for both the turf and the trees, each will be under less stress, minimizing the health care problems.

The Plant Health Care Philosophy

Plant Health Care (PHC) is a comprehensive program that focuses on the plant and the client (in this case, the golf course manager). It is a holistic management program based on management and prevention of plant problems by maintaining the health, appearance, and vitality of the plants in the landscape. Caring for trees on a golf course cannot be viewed separately from the turf care. Trees and turf interact and compete in the same environment, and maintenance practices for each will affect the other. PHC also recognizes that the degree of tolerance for pests, diseases, or other problems varies with the client and the site. As a rule, tolerances on golf courses are quite low.

The key to the success of any PHC program is monitoring the plants. Early detection and correction of situations that might cause plant stress can head off more serious problems in the future. If pest problems, diseases, or other disorders are discovered, the severity and potential impact of the situation must be assessed before deciding on an appropriate response.

Ornamental horticulture has borrowed the concept of *thresholds* from IPM applications of agronomic crops. Where plants are grown as crops, the threshold for treatment of a pest population is determined by doing a cost/benefit analysis of crop loss versus treatment expense. The thresholds for intervention measures on ornamental plants are less quantita-

6.1. Monitoring for symptoms and signs of stress is a critical component of plant health care.

tive and depend on the objectives of the client. As an analogy, the threshold population of spiders in the bathroom that requires immediate intervention is one. Even though spiders are considered beneficial, there is generally no tolerance for their existence in the bathroom. Likewise, although some pest or disease problems have little effect on the overall health of a plant, the aesthetic qualities may be temporarily spoiled. Treatment may be indicated.

Another important aspect of PHC is client education. In the previous example, treatment was called for even though the problem was primarily cosmetic. In some cases it will be preferable not to treat cosmetic problems, but to make "the client" understand the nature of the pest or disease and the minimal impact on the tree's health. This is part of the theory of appropriate response.

Most arborists are now aware that tree health problems are usually the result of several stress factors. Plant management is based on keeping the tree in good health. Recommendations for broad spraying of insecticides or fungicides are much more limited. While pesticides play an important role in plant health management, they also have limitations. In addition to the environmental and health issues associated with the spraying of certain pesticides, pest control may be limited and temporary if other control measures are not integrated into the program.

One undesirable effect of broad insecticide application is that it may reduce the populations of beneficial insects as well as pests. The result could be target pest resurgence of more chemical-resistant pests in the absence of natural predators. Usually there is a natural balance in the ecosystem. In fact, it is not practical to attempt total eradication of any one pest since this could lead to the demise of its predators or parasites. Any reintroduction of the pest could lead to a repeated and perhaps more severe occurrence of the pest's attack.

By focusing on the overall health of the plant rather than the presence of particular pests or diseases, the plant health care provider can identify underlying causes of plant stress. If sources of stress are reduced, other plant health problems will decrease.

Tree Stress

The basic factors that promote plant health include sufficient water and oxygen, optimum temperature and light, and a proper balance of nutrients. Too much or too little of any of these elements can increase stress. "Stress" is a term used to describe any condition that causes a decline in tree health. Stress can be a cumulative problem. Initial stress factors can lower the resistance of the tree, predisposing it to secondary problems. Certain insects are actually attracted to chemical signals produced by a plant "in trouble." As the stress factors increase, the tree enters a "mortality spiral" of declining health.

Stress may be classified into two broad categories: acute and chronic. Acute stress, which can be caused by such factors as improper pesticide sprays or untimely frosts or freezes, occurs suddenly and causes almost immediate injury. Chronic stress takes a longer time to affect plant health, and may be the result of such factors as nutritional imbalance, improper soil pH, long-term weather changes, incorrect light intensity, or others.

Early signs of stress might include reduced growth rate, abnormal foliage color, vigorous epicormic sprouting, or premature leaf drop. The most common causes of tree stress are environment related. If a tree is not well suited for the site in which it has been planted, it is more likely to become stressed. Frequent stress problems include excess or inadequate water, extreme cold, soil conditions such as compaction, or mechanical injury from construction damage.

**6.2. One example of a stress-related pest problem is bark beetles.
They are often the final nail in the coffin of a declining tree.**

Stress is not necessarily an irreversible condition. If the stress factor can be identified, it is sometimes possible to alleviate the condition and improve the health of the tree. However, if only the secondary problems are identified and treated, it is likely that the tree will continue to decline.

Diagnosing plant problems requires a combination of knowledge, experience, and keen observation. Most often it is not simply a case of identifying an insect or disease. Trees that die or decline usually are suffering from a combination of stress factors, and insects or diseases are often secondary, attacking the weakened tree. It takes some detective work to piece together all of the clues. Arriving at a diagnosis is often only the first step. The important second step is to make appropriate recommendations.

Diagnosing Tree Health Problems

The biggest obstacle in diagnosing tree problems is the lack of available information. Obviously, the tree cannot describe the symptoms;

6.3. Flooding can be a major stress for trees. Eventually, oxygen-dependent respiration will halt, and the trees will begin to decline. Flooding can be either an acute stress, or a chronic stress caused by poor drainage or overirrigation.

thus, it is up to the diagnostician to uncover the answers. Although turf usually expresses symptoms of stress within days or weeks, it may take trees years to show signs of trouble. The history of the tree is of utmost importance. How long has the problem been going on? What were the early symptoms? Has there been any construction, excavation, or chemical treatment in the area?

Correct diagnosis of tree problems requires a careful examination of the situation and systematic elimination of possibilities by following a few important steps.

1. **Understand the client's concerns.** What is perceived as a problem by one client may not be by another. Just as you should

6.4. Diagnosis involves careful inspection of the tree and site, as well as gathering information about past occurrences.

not guess what is the problem with the plant, you should not assume what is a concern for the client.

2. **Accurately identify the plant.** Many insects and diseases are specific to certain host trees. Knowing the identity of the plant can quickly limit the number of suspected causes.

3. **Look for a pattern of abnormality**. You must first know what is normal for that plant before being able to recognize the abnormal. For example, when some species of pines shed their interior needles in the fall, some people become concerned that their trees are dying. It may also be helpful to compare the affected tree with other plants on the site, especially those of the same species. Differences in foliage color or growth may present clues as to the source of the problem. Nonuniform damage patterns can be indicative of living factors such as insects or pathogens. Uniform damage over a large area (perhaps several plant species) usually indicates nonliving factors such as mechanical injury, chemicals, weather, and many others.

4. **Carefully examine the site.** Check the contour of the land and note the structures present. If affected plants are restricted to a walkway, road, or fence, the disorder could be a result of wood preservatives, deicing salts, or other harsh chemicals. The history of the property may reveal many situations such as grade changes, excavations, herbicide applications, or gas leaks. Examine the area where the tree is located. How do plants of the same species look?

5. **Note the color, size, and thickness of the foliage.** Dead leaves at the top of the tree are usually the result of mechanical or environmental root stress. Twisted or curled leaves may indicate viral infection, insect feeding, or exposure to herbicides. Early autumn color can be a sign of girdling roots or other root related problems.

6. **Check the trunk and branches.** Examine the trunk thoroughly for wounds as they provide entrances for canker and wood rotting organisms. Small holes may indicate the presence of borers, bark beetles, or sapsuckers. Epicormic shoots may be an indication of stress.

7. **Examine the roots.** Note the color of the roots. Healthy, fibrous roots are generally white and fleshy. Brown roots may indicate dry soil conditions or the presence of toxic chemicals.

Black roots often reflect overly wet soil or the presence of root-rotting organisms.

When diagnosing a tree problem, begin by trying to systematically rule out certain possibilities. The majority of tree health problems are not caused by insects, mites, fungi, or bacteria. Rather, 70 to 90 percent of all plant problems result from adverse cultural and environmental conditions such as soil compaction, drought, moisture fluctuations, temperature extremes, mechanical injuries, or poor species selection.

Diagnostic ability is a cumulative skill. The best diagnosticians learn every time they look at trees. Always gather as much information as possible before committing to a diagnosis. Keep an open mind. As an example, the arborist who finds wood decay, cankers, or borers and dwells exclusively on the fungus or insects rather than the reason for their presence, may be concentrating more on the symptoms than the cause. Many insects and fungi are secondary and only become a problem on stressed trees.

Symptoms and Signs

In diagnosing tree problems, arborists must look for symptoms and signs to determine the cause. Symptoms are effects of the injury or disease that are apparent on the tree. Examples of symptoms include chlorosis, wilting, and leaf scorch. Rarely can a problem be diagnosed by a single symptom. Wilting, for example, can be the result of drought, root problems, or various fungal or bacterial organisms. Signs are direct indications of causal agent problems. Signs might include insect frass, emer-

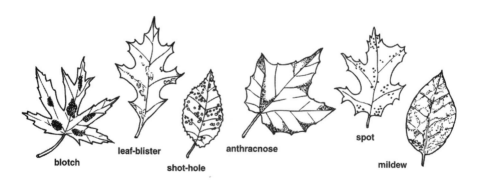

6.5. Patterns of damage on the foliage can indicate the possible causes.

gence holes, or discarded skins. Some of the more common tree ailment symptoms are defined below.

- **Leaf spot**—spots of dead tissue on the foliage; the size, shape, and color may vary with the causal agent, but are usually limited to a small portion of the leaf.
- **Leaf blotch**—dead areas of tissue on the foliage, irregular in shape and larger than leaf spots.
- **Blight**—dieback of major portions of a tree, especially the young, growing tissues.
- **Scorch**—browning and death of indefinite areas along the leaf margins and between the veins.
- **Wilt**—drooping of leaves or shoots, often due to lack of water in the tree.
- **Canker**—localized dead stem tissue, often shrunken and discolored.
- **Stunting**—reduced plant growth.
- **Gummosis**—exudation of sap or gum from wounds or other openings in the bark.
- **Rust**—orange or reddish-brown pustules on leaves, or galls and cankers on stems caused by certain fungi.
- **Gall**—swollen plant tissue that may be induced by insects, fungi, bacteria, or nematodes.
- **Chlorosis**—yellowing of normally green tissues due to a lack of chlorophyll.
- **Necrosis**—death of tissue.
- **Dieback**—large dead portions in the tree.
- **Powdery mildew**—white or grayish fungal growth on the surface of plant tissues, usually leaves.
- **Downy mildew**—fungal growth emerging from beneath the leaf surface.
- **Vascular discoloration**—darkening of the wood's vascular elements.
- **Witches' broom**—abnormal development of multiple secondary shoots, forming a broom-like effect

Many of the symptoms observed in a declining tree are nonspecific. Often they do not lead to a direct diagnosis since there can be many causes. Nonspecific symptoms must be analyzed in combination with specific information about the tree and the site in forming a diagnosis.

Abiotic Disorders

Abiotic disorders are caused by nonliving agents. In diagnosis, it is important to be particularly aware of possible noninfectious disorders. Physiological disorders account for a large percentage of tree deaths. Physiological disorders are the result of noninfectious agents that affect the normal growth and development of the tree. Many agents of physiological disorders are primary stress factors. Some examples include insufficient or excess water, soil and root problems, girdling roots, and mineral deficiencies or toxicities. Other physiological disorders result from improper tree selection for the site, and improper planting or care. Some physiological problems can also be related to weather extremes.

Physical and Mechanical Injuries

Physical or mechanical injuries differ from physiological disorders in that physiological disorders usually occur over a long period of time, while mechanical injuries are relatively sudden. Examples of injuries include fire injury, animal feeding, lightning, vandalism, trimmer and lawn mower damage.

Often the full extent of damage due to physical injuries cannot be immediately assessed. In some cases, initial treatment should be limited until the extent of tree damage can be determined. If the bark is significantly damaged, loose bark should be removed, taking care not to damage healthy, intact tissues.

Sometimes the causes of physical damage can be eliminated to prevent injury to the tree. For example, where trimmer or lawn mower injury is a problem, the area around the tree should be mulched so that grass will not need to be trimmed away from the trunks. The mulch will also reduce moisture stress and competition with the turf. Young trees will benefit from the added protection of perforated PVC pipe.

Soil and Site Problems

A common mistake in diagnosis is to carefully examine the trunk and crown of the tree while overlooking the root condition. Root related problems are often difficult to diagnose due to their limited accessibility. Often symptoms observed in the upper portions of a tree result from poor root health. Hundreds of things can cause root health problems, and the symptoms may be similar for many.

One common site problem is compacted soil. Without sufficient pore space, water and particularly oxygen will be limited. Roots may not be able to penetrate the compacted soil. Root growth and water absorption will decrease, causing the tree to decline or die. A related problem is soil grade increases, which have a similar effect.

If a transplanted tree declines or dies within the first year, the most likely cause is either a lack or an excess of water. If the tree's root ball is too small, or a saturated condition restricts new root formation or water absorption, it will suffer not only from the lack of water, but also from mineral deficiencies.

Competition and Allelopathy

Trees often decline as a result of competition with other trees and plants. The most common example is competition with other trees for sunlight. Trees growing under the canopy of larger trees may exhibit lower branch dieback, stem curvature, and reduced growth. Another

6.6. Chlorosis is one symptom of nutrient deficiency, often resulting from a poor match of the species' ecological requirements to the site conditions.

form of competition takes place below the soil line. Trees may compete with turf and other plantings for available soil nutrients, water, and growing space.

Allelopathy is the chemical inhibition of growth and development of one plant by another. Many trees and other plants produce chemical substances that affect the growth of other plants. Allelopathic chemicals may be exuded directly from the plant or released indirectly through decomposition. These chemicals may inhibit growth, seed germination, flowering, or fruiting of nearby plants. In most cases the inhibition is minor and not easily diagnosed. Stressed plants are more susceptible to allelopathic effects than nonstressed plants.

A few trees are considered highly allelopathic (walnuts and sugar maple, for example). The classic example is black walnut, *Juglans nigra*, which produces a toxin called juglone. This substance stunts the growth of other plants in the area. Many plants will not grow at all in the vicinity of a black walnut tree.

Young trees are most seriously affected by allelopathic chemicals. Tall fescue and other grasses can markedly retard the growth of young trees unless the turf is kept at least two feet from the trunk. The allelopathic effects of turf are usually less of a problem once a tree has an extensive root system.

Chemical Injury

Although any number of chemicals can kill or injure a tree, herbicides are the most frequent culprits. Since herbicides are formulated to kill plants, they are the most phytotoxic of the pesticides used on the golf course. Herbicides, such as glyphosate used to kill broadleaf weeds, may harm nontarget broadleaf plants such as trees and shrubs. These materials are systemic, which means they move throughout the plant. Exposure may be a result of; (1) drift, (2) evaporation of the chemical from heated surfaces surrounding the tree (e.g., solar radiation on tarmac), or (3) accidental spraying on basal or epicormic growth or absorption through thin bark.

If a tree has been exposed to these materials, the leaves will often begin to curl and cup, and the shoot tips may become twisted. The plant may appear wilted. The foliage may become yellowish before dying. Exposure to other herbicides may cause trees to exhibit similar symptoms. Additional symptoms of systemic herbicide injury include veinal or interveinal chlorosis, marginal chlorosis, and leaf fall.

To minimize chances of accidental herbicide injury, never use the same spray equipment to apply herbicides that is used for fungicides or insecticides. Spray only on cool, calm days, and avoid drift. Using low pressure increases droplet size and reduces drift. Apply the herbicide only to target plants, and always check the product label for possible phytotoxicity to the target or surrounding plants.

Insects and Other Pests

At any given time there may be numerous species of insects present on a tree. Few insects are harmful to any one species, though almost every species of tree has at least one insect pest that can cause some problems. Many insects are predators or parasites of harmful insects. Insects have complex life cycles; one stage may cause problems while the next does not. Knowledge of pests' life cycles is important in identification and treatment. Plant health care providers must be able to distinguish both harmful and beneficial insects, and know the degree of damage the plant can tolerate before any control is necessary.

Most insect damage to trees is either a result of feeding or egg laying activity. Insect feeding damage is characterized by the type of mouthparts that the insect has. Insects such as caterpillars, webworms, beetles, and weevils chew on plant parts. Chewing insects eat plant tissue such as leaves, flowers, buds, and twigs. An indication of damage by these insects is often seen as uneven or broken margins on the leaves or other affected plant parts. Others eat only the interveinal tissue, creating a skeletonized leaf. Leaf miners feed between the leaf surfaces, hollowing out tunnels inside the leaves. Some chewing insects eat the leaves completely.

Borers are chewing insect larvae that tunnel under the bark and often into the wood of trees. Because each kind has its own style and tunnel pattern, borers may be identified by their work even after they have left the scene. Trees infested with borers typically show a thinness of crown and a decline in vitality. Conclusive symptoms are small emergence holes in the trunk or branches with frass (semidigested wood). Borers eat the inner bark, phloem, and cambium, thereby interfering with the transport of photosynthate.

Other insects feed by piercing and sucking. Aphids, scales, leafhoppers, and true bugs feed by piercing the plant vascular tissue and cells and sucking out the contents. Symptoms of this type of feeding include chlorosis, stippling, drooping, and distortion.

Mites are close relatives of insects but are actually arachnids as are spiders and ticks. They are very tiny and a hand lens is usually required for identification. Eriophyid mites often cause galls to form on foliage and twigs as a result of feeding or egg laying. Spider mites cause a stippling or bronzing of the foliage as a result of their feeding.

Some insects are vectors of plant diseases. This means that they carry a pathogen, or disease-causing organism, from tree to tree. Dutch elm disease and oak wilt are examples of fungal pathogens that are often spread by bark beetles. Bees can spread fireblight bacteria when collecting nectar from flowers. Aphids and leafhoppers can transmit viruses.

Another group of common plant pests is the nematodes. Nematodes are microscopic worms. Hundreds of species of nematodes have been identified. Not all nematodes are parasites of plants. Some are parasites in insects. Most, however, are free-living in the soil and are important promoters of organic decay. Pathogenic nematodes usually enter the tree through roots, wounds, stomata, or even directly through plant cells. Nematode feeding can cause swelling, deformation of the leaves, blockage of vascular tissue, and even death of the tree. Some nematodes are vectors (carriers) of pathogens.

**6.7. Borers, such as bronze birch borer,
attack trees that are already stressed.**

Diseases

Three requirements are necessary for a tree disease to become serious: a susceptible host, a pathogenic organism, and an environment suitable for disease development. Some pathologists list a fourth element, time, since the other required elements must be present at the same time. Most pathogens are host-specific, meaning they attack a specific plant species or genus. A few pathogens attack a broad range of trees.

Often the part of the tree that is affected is an indication of the severity of the disease. For example, diseases that affect only the foliage may not be a major problem unless defoliation occurs in several consecutive years inducing stress to the tree. However, vascular diseases such as oak wilt and Dutch elm disease tend to be fatal.

Fungi cause most plant diseases. However, that does not mean that most fungi cause disease. Only a small percentage of fungi cause disease in plants. Most are beneficial. Most trees are susceptible to infection by at least one disease-causing fungus. The severity and extent of the dis-

6.8. Many leaf-feeding insects, such as the eastern tent caterpillar, are more of a nuisance and an aesthetic problem than they are a threat to the tree.

ease depends upon the resistance and vitality of the plant, the virulence of the organism, and environmental conditions.

Besides fungi there are many other disease causing organisms or pathogens. Two common diseases, fireblight and crown gall, are caused by bacteria. Plant tissues that are infected with bacteria frequently appear water-soaked and may have a foul odor. Other diseases such as ringspot, yellowing, and some stunt diseases are caused by viruses and phytoplasmas.

Control of plant diseases may be considered an exercise in integrating pest management tactics. Most fungicides are best used to prevent disease spread rather than to eliminate existing disease problems. Steps that are taken to maintain tree health and improve cultural conditions are the best disease prevention measures.

Using a Diverse Management Arsenal

Integrated pest management (IPM) is a systematic approach to insect and disease management. IPM is an important component of Plant Health Care. Management within a plant health care program should

6.9. Diseases, such as anthracnose, that affect primarily the foliage of trees are usually not life-threatening.

incorporate a combination of strategies including resistant plants, cultural management, mechanical controls, and chemical management of plant pest problems. The goal is to maintain tree health while minimizing any adverse ecological impact from management tactics.

One important cultural management technique is to select trees that are resistant to known insects or diseases. Avoid planting trees in sites for which they are ill suited. Subjecting a tree to stress predisposes it to other health problems.

Chemical management is a valuable option that should be kept in the arsenal. Insecticides, fungicides, and herbicides play an important part in the health appearance and vitality of the trees, turf, and other plants on a golf course. Superintendents must recognize, however, that broadcast sprays of insecticides are rarely the answer. Target application to specific pests at the proper time can be a better way to manage plant health. In addition, the use of horticultural oils, insecticidal soaps, and alternative insecticides can be an effective way to manage pest populations while minimizing adverse affects to natural enemies.

Consider alternatives to chemical control of pests. As mentioned previously, many insects live in a natural balance with predators and parasites. Biological management involves the manipulation or conservation of natural pest predators and parasites to maintain pest populations at levels within tolerance. In some cases where the pest population has far exceeded the predator population, the predators have been artificially reintroduced to establish the balance.

Another biological control measure is to incorporate natural toxins or diseases of the pests. Two examples are *Bacillus thuringiensis* (Bt) and *Bacillus popilliae,* which produce toxins that are fatal in certain insect populations. Extracts of these bacteria are sold commercially. Some biological controls require ideal conditions to become established and can be slow in providing initial control.

The use of insect hormones is another biological control method. One example is the use of juvenile hormones to prevent insects from reaching sexual maturity, which prevents them from reproducing. Insect pheromones are often used to attract male insects to traps to determine population levels and aid in the timing of spraying.

Disease control is often achieved best through a combination of control measures. Environmental and cultural control measures can be very important. For example, if a tree is susceptible to leaf diseases, it should be planted in a sunny site, and water on the foliage should be minimized. Good sanitation practices such as raking leaves and removing diseased branches will help control many diseases. Pruning to promote rapid drying is an effective control practice for some disease prevention.

Some diseases are controlled by killing the insect vectors. Since insects spread many diseases, controlling the carrier (vector) may reduce the spread of the disease. Chemical control of insects and diseases is still an important part of maintaining tree health. However, if chemicals are to be used, they must be used wisely. Pesticides should be selected with regard to minimizing the effect on nontarget organisms. Spot spraying is preferred to broadcast spraying. All chemical applications must be

6.10. Plant Health Care reduces the need for large-scale spraying of trees. If it is required, however, arborists that are licensed pesticide applicators with the appropriate equipment should be employed to do the job.

made in accordance with applicable laws and safety regulations, and according to label instructions.

The Tree's Natural Defense System

All trees possess certain traits that deter some pests and diseases. Some have thorns, spikes, or tiny hairs on the leaves that are a physical deterrence. Some have a thick, toughened cuticle. The cellulose and lignin in tree cells is indigestible to many insects, animals, and even some pathogens.

Trees also produce allelochemicals such as tannins and phenols that have toxic or deterrent effects on certain insects. In fact, we use extracts of some of these allelochemicals for pesticides. Nicotine, pyrethrin, and rotenone are examples of these natural insecticides. These allelochemicals may be the reason most insects are host-specific. Insects have developed special enzymes that allow them to detoxify specific compounds.

6.11. Of course, some tree "pests" are not influenced by tree stress. Beavers selected these trees because of their proximity to the water. The felling techniques are questionable since the trees are not always felled in the desired direction.

Recent studies have investigated the relationship between cultural influences and allelochemical production in trees. For example, moderate drought stress can increase levels of allelochemicals, perhaps boosting the defenses of the tree. Conversely, sun-adapted trees grown in shade have reduced levels of allelochemicals and lowered resistance.

Much of the research has concentrated on the effects of nitrogen fertilization. Researchers have observed that rapidly growing trees are sometimes less resistant to certain insects and diseases. This is contrary to the traditional belief that a rapidly growing tree is a healthier tree. One explanation is that photosynthate is diverted from defense (production of allelochemicals) to increased growth of succulent tissues. Increased growth without a corresponding increase in photosynthesis may stress the plant.

This research may have significant ramifications for golf course superintendents. Trees located near fairways, tees, and greens receive high levels of nitrogen fertilization. It is important to monitor these trees closely for signs of stress or insect infestation. Heavily fertilized trees may require supplemental irrigation. In addition, more frequent pesticide applications may be necessary.

Summary

1. Trees on golf courses present some intense plant health care scenarios, yet their management and care usually comes second to that of turf. Unfortunately, the turf care and tree care can be mutually detrimental. The challenge for golf course superintendents is to maintain both in good health.
2. Plant Health Care (PHC) is a comprehensive program that focuses on the plant and the client. The goal is to prevent plant problems by maintaining the health, appearance, and vitality of the plants as part of the landscape.
3. Tree health problems are usually a result of several stress factors. "Stress" is a term used to describe any condition that causes a decline in tree health. The most common causes of tree stress are environment related—excess or inadequate water, poor soil conditions, nutritional deficiencies, etc.
4. Diagnosis of plant problems requires a combination of knowledge, experience, and keen observation. Often it takes some detective work to piece together all of the clues. Arriving at a

diagnosis is only the first step, however. The important second step is to make appropriate recommendations.

5. A good PHC monitor must recognize the various signs and symptoms of insects, diseases, and physiological disorders of plants. An understanding of pest and disease life cycles is essential before recommending treatments.

6. Plant Health Care depends on a diverse management arsenal. Although many stress factors can be avoided through careful planning, plant selection, and thoughtful management, often the mistakes have been made long before the management professional becomes involved. Thus, management alternatives are often limited, and treatments are sometimes marginally effective.

7. Recent research has focused on trees' natural defense systems. The information gained is shedding new light on the concept of Plant Health Care. Research on the relation between fertilization and insect problems, for example, may have a significant impact on golf course superintendents' maintenance methods.

7

Tree Maintenance

It is common to consider trees to be a low maintenance feature of the golf course landscape. Compared to turf, flowers, and some shrubs, trees require relatively little maintenance time. Low maintenance, however, is not synonymous with neglect. Golf course trees require occasional pruning, periodic fertilization, and at times, structural support. The consequences of failure to maintain the trees on a golf course range from reduced functional values, such as screening between fairways, to tree failures.

Pruning

Pruning is the most common tree maintenance procedure. Although forest trees grow quite well without pruning, golf course trees require a higher level of care to maintain their safety, function, and aesthetic appeal. Pruning should be done to achieve a desired response. Improper pruning can cause damage that will last for the life of the tree, or worse, shorten the tree's life.

Reasons for Pruning

Since each cut has the potential to change the growth of the tree, no branch should be removed without a reason. Common reasons for pruning are to remove dead branches, to improve tree structure, to provide clearance for buildings or people, and to eliminate hazards. Trees may also be pruned to increase light and air penetration to the inside of the tree's crown or to the landscape or turf below. On golf courses, mature trees are pruned as a corrective measure, to provide clearance for play, or to increase sunlight.

Routine thinning does not necessarily improve the health of a tree. Trees produce a dense crown of leaves to manufacture the sugar used as energy for growth and development. Removal of foliage through pruning reduces photosynthesis and thus reduces growth and stored energy reserves. Heavy pruning can be a significant health stress for the tree.

If people and trees are to coexist in the manicured, aesthetic environments of golf courses, then we sometimes have to modify the trees. These environments do not mimic natural forest conditions. Safety is a major concern since players and staff members are in close contact with the trees. Also, golf course turf requires a great deal of sunlight, so shade must be controlled. Proper pruning, with an understanding of tree biology, can maintain good tree health and structure while enhancing the aesthetic and economic values of the course.

When to Prune

Most routine pruning to remove weak, diseased, or dead limbs can be accomplished at any time during the year with little adverse effect on the tree. As a rule, growth is maximized and wound closure is fastest if pruning takes place before the spring growth flush. Some trees, such as maples and birches, tend to "bleed" if pruned early in the spring. This may be unsightly, but is of little consequence to the tree.

A few tree diseases, such as oak wilt, can be spread when pruning wounds allow spores access into the tree. Susceptible trees should not be pruned during active transmission periods.

Heavy pruning just after the spring growth flush should be avoided. This is when trees have just expended a great deal of energy to produce foliage and early shoot growth. Removal of a large percentage of foliage at this time can stress the tree.

Making Proper Pruning Cuts

Pruning cuts should be made just outside the branch collar. The branch collar contains trunk or parent branch tissue and should not be damaged or removed. If the trunk collar has grown out on a dead limb to be removed, make the cut just beyond the collar. Do not cut the collar.

If a large limb is to be removed, its weight should first be reduced. Making an undercut about 12–18 inches from the limb's point of attachment reduces the potential for bark tearing. A second cut is made

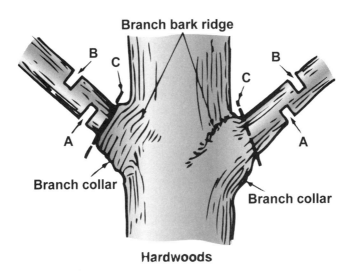

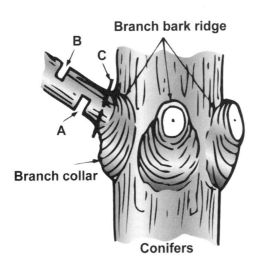

7.1. **Pruning principles. The first cut (A) undercuts the limb. The second cut (B) removes the limb. The third cut (C) should be just outside the branch collar to remove the resultant stub.**

from the top, a few inches farther out on the limb. This removes the limb, leaving the 12–18 inch stub. The stub is removed by cutting back to the branch collar.

Pruning Techniques

Specific types of pruning may be necessary to maintain a mature tree in a healthy, safe, and attractive condition.

- *Crown cleaning* is the removal of dead, dying, diseased, crowded, weakly attached, and low-vigor branches from the crown of a tree.
- *Crown thinning* is the selective removal of branches to increase light penetration and air movement through the crown. Thinning opens the foliage of a tree, reduces weight on heavy limbs, and helps retain the tree's natural shape.

7.2. Cuts should be made to the branch collar. Leaving a stub inhibits proper closure of the wound.

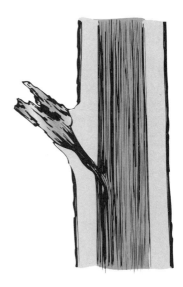

7.3. Leaving a stub maintains an open pathway to decay.

- *Crown raising* removes the lower branches from a tree in order to provide clearance for buildings, vehicles, pedestrians, and vistas. Crown raising can be a very effective method of increasing light penetration and air circulation to turf below trees.
- *Crown reduction* reduces the size of a tree, often for clearance for utility lines. Reducing the height or spread of a tree is best accomplished by pruning back the leaders and branch terminals to lateral branches that are large enough to assume the terminal roles (at least one-third the diameter of the cut stem). Compared to topping, this helps maintain the form and structural integrity of the tree.

A Few Words about Topping

Topping is the indiscriminate cutting back of tree branches to stubs or lateral branches that are not large enough to assume the terminal role. Other names for topping include "heading," "tipping," "hat-racking," and "rounding over." The most common reason given for topping is to reduce the size of a tree. People fear that tall trees may pose a

7.4. A properly made pruning cut should close completely, and compartmentalize internally.

hazard. Topping, however, is not a viable method of height reduction, and certainly does not reduce the hazard. In fact, topping will make a tree more hazardous in the long term.

Topping often removes 50–100 percent of the leaf-bearing crown of a tree. Since the leaves are the "food factories" of a tree, this can temporarily "starve" a tree. The severity of the pruning triggers a sort of survival mechanism. The tree activates latent buds, forcing the rapid growth of multiple shoots below each cut. The tree needs to put out a new crop of leaves as soon as possible. If a tree does not have the stored energy reserves to do this, it will be seriously weakened and may die.

7.5. Pruning cuts made at the correct place, just beyond the branch collar, will form a complete "doughnut" ring of woundwood as they close. Avoid making large cuts in close proximity because internal decayed/discolored wood may coalesce.

7.6. Severe pruning can lead to sunscald on exposed branches and profuse watersprout development.

7.7. Stubs left in topped trees will often decay, creating an unsightly condition and a potential hazard.

The survival mechanism that causes a tree to produce multiple shoots below each topping cut comes at great expense to the tree. These shoots develop from buds near the surface of the old branches. Unlike normal branches that develop in a "socket" of overlapping wood tissues, these new shoots are only anchored in the outermost layers of the parent branches. Unfortunately, the shoots are very prone to breaking, especially during windy conditions. This is a potential liability for the golf course owners and the superintendent. Since topping is considered to be an unacceptable pruning practice, any damage caused by branch failure of a topped tree may lead to a finding of negligence in a court of law.

How Much Should Be Pruned?

The amount of live tissue that should be removed depends on the pruning objectives as well as tree size, species, and age. Younger trees will tolerate the removal of a higher percentage of living tissue than mature trees. An important principle to remember is that a tree can recover from several small pruning wounds faster than from one large wound.

7.8. Topping trees ruins their natural beauty and can make them prone to branch failure.

A common mistake is to remove too much inner foliage and small branches. It is important to maintain an even distribution of foliage along large limbs and in the lower portion of the crown. Overthinning reduces the tree's sugar production capacity and can create tip-heavy limbs that are prone to failure. In addition, overthinning often leads to excessive sucker (watersprout) growth that is less desirable than the original, natural growth. Keep in mind that the trees will eventually grow into voids left by pruning, requiring additional pruning or removal later.

Mature trees should require little routine pruning. A widely accepted rule of thumb is never to remove more than one-fourth of a tree's leaf-bearing crown. In a mature tree, pruning even that much could have negative effects. Removing even a single, large-diameter limb can create a wound that the tree may not be able to close. The older and larger a tree becomes, the less energy it has in reserve to close wounds and defend against decay or insect attack. The pruning of large, mature trees is usually limited to the removal of dead or potentially hazardous limbs.

Establishing a Strong Scaffold Structure

A good structure of primary, scaffold branches should be established while the tree is young. The scaffold branches provide the framework of the mature tree. Properly trained young trees will develop a strong structure that will require less corrective pruning as they mature.

The goal in training young trees is to establish a strong trunk with sturdy, well-spaced branches. The strength of the branch structure depends on the relative sizes of the branches, the branch angles, and the spacing of the limbs. Naturally, this will vary with the growth habit of the tree. Pin oaks and sweetgums, for example, have a conical shape with a central leader. Elms and live oaks are often wide-spreading without a central leader. Other trees, such as lindens and Bradford pears, are densely branched. Improve tree structure by pruning to remove structurally weak branches while maintaining the natural form of the tree.

Trunk Development

For most young trees, maintain a single, dominant leader. Do not prune back the tip of this leader. Do not allow secondary branches to outgrow the leader. Sometimes a tree will develop double leaders known as codominant stems. These can lead to structural weaknesses, so it is best to remove one while the tree is young.

The lateral branches contribute to the development of a sturdy, well-tapered trunk. It is important to leave some of these lateral branches in place, even though they may be pruned out later. These branches, known as temporary branches, also help protect the trunk from sun and mechanical injury. Temporary branches should be kept short enough not to be an obstruction or compete with selected permanent branches.

7.9. Codominant stems, if allowed to develop as the tree matures, will be a likely point of failure in the future.

Permanent Branch Selection

Nursery trees often have low branches that may make the tree appear well-proportioned when young, but low branches are seldom appropriate for large growing trees in a golf course environment. How a young tree is trained depends on its primary function in the landscape. For example, trees that line the golf cart path must be pruned so that they allow at least 10 feet of overhead clearance.

The tree's intended function and location on the course determine the height of the lowest permanent branch. Trees that are used to screen

7.10. A tree that has been properly pruned when young will develop a strong scaffold structure, with branches well-spaced vertically and radially along the trunk.

a view or provide a windbreak may be allowed to branch low to the ground. Trees near a tee box may be pruned high to provide adequate sunlight to the turf.

The spacing of branches, both vertically and radially in the tree is very important. Branches selected as permanent, scaffold branches must be well-spaced along the trunk. Maintain radial balance with branches growing outward in each direction.

A good rule of thumb for the vertical spacing of permanent branches is to maintain a distance equal to 3 percent of the tree's eventual height.

7.11. Poor pruning cuts made flush with the trunk may not close properly. Pruning paint does not speed closure or prevent decay.

Thus a tree that will be 50 feet tall should have permanent scaffold branches spaced about 18 inches apart along the trunk. Avoid allowing two scaffold branches to arise one above the other on the same side of the tree.

Some trees have a tendency to develop branches with narrow angles of attachment and tight crotches. As the tree grows, bark can become enclosed deep within the crotch between the branch and the trunk. This is called included bark. Included bark weakens the attachment of the branch to the trunk and can lead to branch failure when the tree matures. You should prune branches with weak attachments while they are young.

Newly Planted Trees

Pruning of newly planted trees should be limited to corrective pruning. Remove torn or broken branches. Save other pruning measures for the second or third year.

The belief that trees should be pruned when planted to compensate for root loss is misguided. Trees need their leaves and shoot tips to provide food and the substances which stimulate new root production. Unpruned trees establish faster, with a stronger root system, than trees pruned at the time of planting.

Wound Dressings

Wound dressings were once thought to accelerate wound closure, protect against insects and diseases, and reduce decay. However, research has shown that dressings do not reduce decay or speed closure, and rarely prevent insect or disease infestations. Most experts recommend that wound dressings not be used. If a dressing must be used for cosmetic purposes, then only a thin coating of a nontoxic material should be applied.

Fertilization

Trees require certain essential elements to function and grow. An essential element is a chemical element that is involved in the metabolism of the tree, or necessary for the tree to complete its life cycle. For trees growing in a forest site these elements are usually present in sufficient quantities in the soil. Golf course trees may be growing in sites that have been modified a great deal. Soils may be stripped of the organic layer and topsoil, leaving an almost sterile subsoil. Other areas may be heavily supplemented with nitrogen to encourage vigorous turf growth.

Fertilizing a tree can increase growth, reduce susceptibility to certain diseases and pests, and under certain circumstances, can even help reverse declining health. If the fertilizer is not prescribed and applied wisely, however, it may not benefit the tree at all and may adversely affect the tree. Mature trees making satisfactory growth may not require fertilization. Nitrogen can unnecessarily increase growth, requiring more pruning. It is important to recognize when a tree needs supplemental fertilizer, what nutrients are needed, and when and how it should be applied.

Tree Requirements

Trees take up essential elements dissolved in water through the roots. Each element has a specific role in the tree and cannot be replaced by another. Certain elements, known as macronutrients, are required in relatively large quantity. The most important of these macronutrients is nitrogen (N), because it is most often limiting. Nitrogen is a constituent of proteins and chlorophyll. If nitrogen is deficient, shoot growth is reduced, and the tree's leaves will turn yellow due to reduced levels of chlorophyll. In addition to nitrogen, phosphorus (P) and potassium (K) are also required in relatively large quantities. The other macronutrients include magnesium (Mg), sulfur (S), and calcium (Ca).

Other elements, known as micronutrients, are required in lesser quantities. Although these elements are not required in large amounts, a deficiency of any one can have profound effects on the health of the tree. For example, iron chlorosis is a condition that results when a tree is not absorbing sufficient quantities of iron. It can eventually kill a tree. Like iron, manganese and zinc may at times be deficient in a tree. Chlorosis symptoms due to manganese deficiency are similar to iron deficiency. Trees that are deficient in zinc may have small leaves and stems that fail to elongate (rosetting). The remaining four micronutrients, molybdenum, copper, chlorine, and boron are less likely to be deficient.

Soil and Foliar Analysis

The most accurate way to determine a tree's nutrient needs is to obtain laboratory analyses of the soil and foliage (leaves). A soil analysis provides information about the availability of essential elements, soil pH, and cation exchange capacity (ability of the soil to hold essential minerals). The pH and salt content (especially in arid regions) are very important for recommending treatments. Some micronutrients, such as iron and manganese, become unavailable when the soil pH is high.

A soil analysis has greater value when done in conjunction with a foliar analysis. Since the roots of trees grow over a vast area and at different depths, the soil analysis alone may not yield accurate results for all soil nutrients. Leaf samples taken from all over the tree, dried and analyzed, may help diagnose certain deficiencies or toxicities. However, a soil analysis or foliar analysis alone can be misleading. Certain minerals may be deficient in the leaves, yet plentiful in the soil and unavailable due to soil pH.

Rates

The rate of fertilizer application depends on the vigor of the tree, the form of the fertilizer, the soil and site conditions, and the method of application. The general recommendation is 2–4 pounds of actual nitrogen per 1000 square feet of root area fertilized. If the fertilizer is being broadcast over turf or ground cover, this rate should be reduced. If a slow-release form of nitrogen is used, a higher application rate (up to 6 pounds per 1000 square feet) can be used. In a golf course setting, many trees may be benefiting from fertilizer applications to the turf. Applications to trees should be made only if indicated by soil/foliar analyses.

Timing

The most limiting factor in fertilizer uptake is water availability. Fertilization studies on mature trees show the response is greatest when moisture levels are high. The frequency of application depends on the tree and the soil conditions. In sandy soils, smaller amounts of fertilizer may need to be applied more frequently than in clay soils because of a greater possibility of leaching. Fertilizer uptake is greatest during periods of active root growth, so applications are most effective in spring and fall.

Fertilizer Types and Forms

Fertilizers are available in organic and inorganic forms. An advantage of organic fertilizers is that they must be converted to inorganic ions before absorption and are not leached as readily from the soil. The solubility of inorganic fertilizers, however, is less affected by temperature so that the rate of availability is more uniform.

Fertilizers are also categorized by the rate they release nitrogen into the soil. Controlled release or slow-release fertilizers release the nitrogen over an extended period of time. This may be controlled by a slowly soluble coating or by nitrogen formulations that are not soluble in water.

Slow-release fertilizers have several advantages. Applications need not be made as frequently. Less nitrogen is lost through leaching. There is less chance for salt buildup in the soil, causing fertilizer "burn." For long-term release of nitrogen, trees should be fertilized with either a slowly soluble source of at least 50 percent water insoluble nitrogen (WIN) or a slow release source with a 20–30 percent dissolution rate.

Application Techniques

Application of fertilizer to the soil surface is the easiest and least expensive method of fertilizing trees. The fertilizer is broadcast over the soil surface at the recommended rates. One disadvantage of this technique is that turf competes with the trees for uptake of available nutrients.

The drill-hole application method is designed to place granular fertilizer below turfgrass roots. Holes spaced 2–3 feet apart are drilled in the soil throughout the root zone of the tree. Care must be taken not to drill into major roots, irrigation systems, or underground utilities. Holes should be 6–12 inches deep, and the fertilizer should be distributed evenly, based on desired rates. If the holes are too deep, the fertilizer will be below the area of active uptake by the roots. One advantage of drill-hole fertilization is the aeration of the soil. This may not be a viable option along fairways, green, or tees since it disrupts the turf.

A third application technique is the liquid injection method. Fertilizer dissolved or suspended in water is injected into the soil using a lance and hydraulic sprayer. The hole spacing and distribution are similar to the drill-hole method. Since this technique and the drill-hole technique can both trigger small patches of vigorous turf growth around the holes, they are sometimes combined with a surface application.

Foliar application is a technique employed to correct micronutrient deficiencies. For example, chelated iron sprays may be used to provide a rapid, although temporary, treatment of iron chlorosis. This method should not be considered as an adequate means of providing all the necessary mineral elements for adequate tree growth. Micronutrient spray applications are most effective when made just before a period of active growth.

Implants and injections are two techniques employed to introduce chemicals directly into the xylem of trees. They have been used to successfully treat micronutrient deficiencies but should not be considered a cure. Most of these products rely on the transpirational stream for uptake and distribution, so applications are most effective on days when the tree is actively transpiring. Many experts are hesitant to recommend

7.12. Drill-hole fertilization.

implants or injections since chemical toxicity may be severe and damage to the cambium and xylem may result. Although treatments using systemic injections may yield remarkable results, repeated use can cause wounds to coalesce. Use should be limited to situations where other options are not viable.

Salt Problems

Recommended fertilizer rates are based on achieving good tree response while avoiding fertilizer "burn." Fertilizers are actually salts of different chemicals that have nutritional value to plants. Fertilizer burn is the same type of injury that can be caused by deicing salts. Excessive soluble fertilizer in the root zone can actually draw moisture out of the roots. Symptoms range from wilting and marginal burning of the foliage to death of the tree. Many fertilizer labels indicate the salt index. Fertilizers with salt indexes less than 50 are usually considered "safe."

7.13. Liquid injection fertilization.

Leaching

Leaching is the washing out of chemicals down through the soil. Nitrogen and other water soluble elements tend to leach from the soil. Not only does this make them unavailable to the tree, but they can be a pollution problem to groundwater, lakes, and streams. Leaching is a major consideration on some golf courses where soils are sandy and chemical use is heavy. To minimize leaching of fertilizer, apply only the kind and amount of fertilizer needed, use organic or slow release forms, and avoid overirrigating sandy soils.

Other Soil Amendments

There are a number of fertilizer supplements and soil amendments available ranging from organic formulations that include seaweed, to innoculants of the fungi involved in mycorrhizae. Some have shown considerable efficacy, especially on extremely poor soils. More research is needed before wide recommendations can be made. Some arborists use supplemental soil amendments when transplanting large trees or attempting to reverse the decline of stressed trees.

Mechanical Support: Cabling and Bracing

Some trees become crucial to the play of a hole. Loss of a significant tree, such as a mature oak that defines a major dogleg, could mean redesigning the fairway since it could take decades for a replacement tree to reach similar dimensions. Trees of such value can justify extra care and the requisite measures to ensure their survival. Mature trees, however, have two dimensions to survival: health and structural stability. It is possible for a tree to have a healthy, vigorous crown and root system, yet be on the verge of collapse. Golf course superintendents must work with arborists to determine the obvious and latent hazards of trees on the golf course. If structural flaws are found, cabling and bracing are two techniques in an arborist's arsenal that may be employed to preserve the trees.

Cables and braces are installed in trees to provide added support of weak, split, or heavy branches. When used wisely, they may extend the life of a tree or make it safer. Cables and braces cannot be relied on to make all hazardous trees safe, however. Some trees, with extensive decay or massive splits, should be removed entirely. The installation of reinforcing hardware cannot prevent all tree failures, and caretakers must be cautioned against being lulled into a false sense of security.

Whenever hardware such as cable and bracing material is installed in a tree, the tree is wounded. Once wounded, there is the risk of decay advancing into wood that supports or holds the hardware. Therefore the situation must be assessed carefully when deciding whether to install cables, to prune out the potentially hazardous limbs, or to remove the tree.

In determining whether cabling is warranted, the condition, size, species, and value of the tree should be considered. If the root system is not structurally sound or if the tree contains excessive decay, removal may be preferable. The installation of cables and braces in trees requires a general knowledge of tree growth and structure, as well as an understanding of the hardware involved and the proper installation techniques.

Reasons for Installing Cables and Braces

There are a number of structural defects in trees that may warrant the installation of support hardware. Crotches that are beginning to split or are weak due to included bark can often be reinforced successfully with braces and cables. Codominant stems are often good candidates for the same reasons. Heavy limbs that extend over structures can sometimes benefit from the added support of a cable. Other examples include wide-spreading branches or multistemmed trees that may be threatened by ice or snow loading.

Once a tree has been inspected and has been declared a candidate for supporting hardware, the arborist must decide what is needed. Braces, or tree bolts, are installed through weak or split crotches to prevent further splitting. Braces are usually accompanied by one or more cables to provide added support. Cables are also installed to help support heavy limbs, especially those which extend over buildings or areas where people congregate.

Bracing

Bracing is the use of steel rods to provide rigid support. The preferred method of installation is to drill a hole, 1/16-inch diameter larger than the rod, directly through both sides of the crotch to be supported. (This obviously requires a drill bit of sufficient length to go all the way through from one side.) The rod is fed through the tree and bolted on each end using large round or oval washers and nuts. Excess rod should be cut off, and the ends peened to prevent the nut from backing off. Any portion of the rod that is exposed should be treated with rust preventative paint.

Sometimes it is necessary to install more than one bracing rod. Multiple rods provide added strength and reinforcement. If more than one bracing rod is installed, they should be staggered. Vertical alignment of the rods may add to the formation of decay columns in the tree. Whenever practical the rods should be spaced at least 12 inches apart.

Cabling Hardware

It is important to select the appropriate hardware for cabling a tree. Cables, eyebolts, and other cabling hardware come in various sizes and

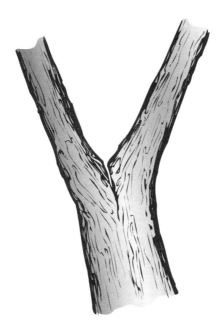

7.14. Splits between codominant branches can sometimes be repaired with braces and cables. If codominant branches are removed when the tree is young, this problem can be avoided in mature trees.

types. Consider the size of the limbs, the weight to be supported, and the presence of decay when choosing materials. If the hardware is too small or inadequate, the cable may fail.

There are two types of cable commonly used in cabling trees, 7-strand common grade galvanized and extra high strength (EHS) cable. The common grade cable is relatively malleable (bendable) and easy to work with. The EHS cable is much stronger but less flexible than common grade cable. Both are available in a wide range of sizes.

Cables can be anchored in trees using any of three types of hardware: lag hooks, eyebolts, or threaded rods with amon eye nuts. A lag hook is a threaded device in the shape of a "J." Lag hooks come with right- and

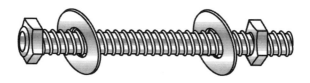

7.15. Bracing rod with washers and bolts.

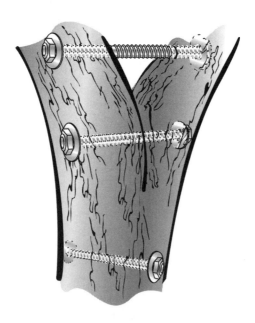

7.16. If multiple bracing rods must be installed they should be at least 12 inches apart, if practical, and not in the same plane.

left-handed threads so that when each end is twisted into the branch to tension the cable, the cable will not come unlaid (unwound).

Lag hooks are installed by screwing into a predrilled hole that is 1/16-inch smaller in diameter than the lag. The lags should be screwed into the tree so that the J-loop ends up vertical, with the open end just contacting the bark. Care must be taken not to injure the bark. Lag hooks work quite well on small limbs and trees with hard wood, but should not be installed in limbs that are greater than eight inches in diameter. Lag hooks should never be installed in limbs with decay. The decay will limit the holding capacity of the lags and may spread into healthy wood.

In circumstances where lag hooks are not appropriate, eyebolts or threaded rods with amon eye nuts may be used. Both are drop-forged and machine threaded, and their installation is similar. A hole not more than 1/16-inch larger than the hardware is drilled through the limb to be cabled. The eyebolt or threaded rod is installed with a round washer and nut on the outside end. The exposed threads on the end of the eyebolt and both ends of the threaded rod used with the amon eye should be peened to prevent the nut from unscrewing. Drop-forged eyebolts are considered slightly stronger than amon eye nuts used with threaded rods. However, an added advantage of the amon eye system is that the length of the rod can be adjusted easily for any job.

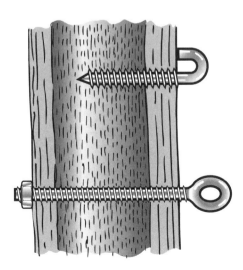

7.17. Installation of a lag screw and an eyebolt.

When attaching the cable to its anchoring hardware, thimbles must be used. The purpose of the thimble is to protect and preserve the cable from excessive wear. If soft common grade cable is installed directly on the hardware, the steel to steel contact and abrasion may eventually cause the cable to break.

Thimbles must also be used when installing Preformed Cable Grips. Preformed Cable Grips are installed over the thimble and then wrapped upon the cable. Cable grips are used to attach EHS cable to the hardware. EHS cable is not malleable enough to form an eyesplice. Cable grips have been found to be quite satisfactory for small and medium-sized trees. However, the cable grips are not designed to withstand dynamic loads that can be created by branches twisting and swaying, particularly in gusting winds. Excessive wind sway in large trees may cause metal fatigue in the Preformed Tree Grips, which could lead to failure.

In order to install the hardware in the tree, holes must be drilled. This can be accomplished using either a brace and bit, or an electric drill. Though fast and efficient, electric drills can be difficult to work with in a tree, and an electrical source is not always available. Battery powered, rechargeable drills offer a solution to these problems. Whether using an electric drill or a brace and bit, the bit should be a ship auger. This will work more efficiently in green wood and will pull the shavings from the hole.

Attaching the Cable to the Hardware

As previously mentioned, if EHS cable is used it must be attached to the hardware using Preformed Cable Grips. If common grade seven-strand cable is used, however, there are alternatives.

One method of attachment is to form an eyesplice. A cable eyesplice is actually a series of wraps and not a true splice. A thimble is always used to form the eye in the end of the cable. The thimble size must match the cable size. If the bend in the thimble is too tight, the cable will be weakened. The eyesplice is made by wrapping the end of the cable around the thimble, then separating the cable strands. Each strand is wrapped individually around the cable with at least two or three turns. When complete, the finished splice will have a neat appearance and will provide optimum holding capacity.

An alternative to forming an eyesplice is to use cable clamps (U-bolts). It is imperative that cable clamps be installed correctly if they are to offer reliable anchorage. If cables are installed, three should be used for adequate strength. They must be installed several inches apart, with the "U" of the clamp over the short end of the cable. Once again, size is important; the clamp size must fit the cable size. Many arborists prefer not to use cable clamps, feeling that eyesplices and Preformed Cable Grips offer more reliable anchorage and a neater appearance.

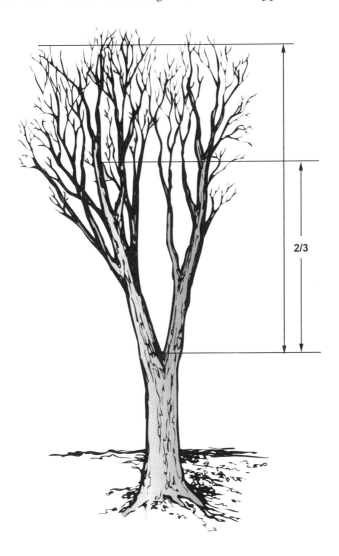

7.18. Cables should be installed at least two-thirds the distance from the crotch to the branch tips.

Cable Installation

Before installing cable in a tree, the tree should be properly pruned. Hazardous limbs should be removed. If necessary, the tree should be trimmed to reduce weight from the ends of long, horizontal limbs.

A general rule of thumb is to install the cable two-thirds the distance from the defect or weak crotch to the ends of the limbs. Exact placement will depend upon the location of lateral branches and defects. The limbs must be large and solid enough at the point of cable attachment to provide adequate support of the hardware.

The angle of the cable and its distance from the crotch determine its strength and effectiveness. Installing the cable directly across the crotch being supported and at least two-thirds up can maximize the support.

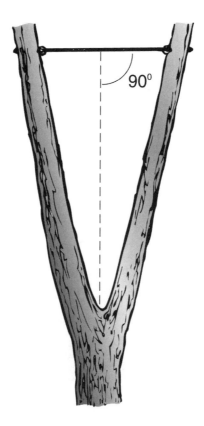

7.19. Cables should be installed perpendicular to (at a 90 degree angle with) an imaginary line that bisects the crotch to be supported.

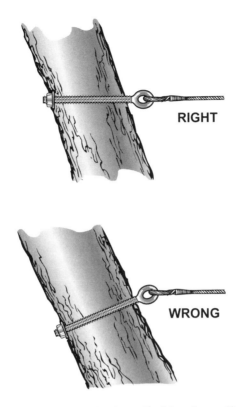

7.20. The eyebolt should be installed in direct line with the
pull of the cable.

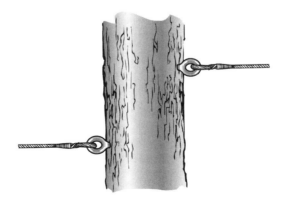

7.21. If multiple cables are to be installed, keep them at
least 12 inches apart, if possible.

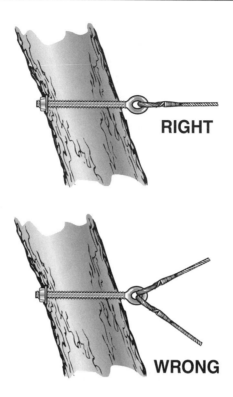

RIGHT

WRONG

7.22. Do not install more than one cable on each eyebolt or lag.

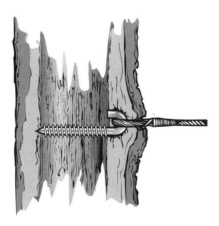

7.23. Most of the holding power comes from new wood
produced after installation.

"Directly across" is determined by setting the cable perpendicular to an imaginary line bisecting the crotch.

The hardware should be installed with the pull of the cable in direct line with the installed bolt or lag. If the cable is installed correctly, the hardware probably will not be installed perpendicular to either branch of attachment. It is important not to have the cable's pull at an angle to the hardware.

Once installed, the cable should be just taut. A cable that is too tight may put excessive strain on the wood fiber, resulting in more damage at the defect or causing the hardware to pull out. The limbs can be brought closer together with ropes or slings and a come-along. This will make the installation easier, and the cable should be taut when the limbs are released. If the cable is installed while the tree is in foliage, it should be tight enough that it will not slacken after the leaves have fallen.

The most common cable installation is the simple or direct cable (one cable between two limbs). Sometimes a tree will require more than one cable. If multiple cables are required, extra strength can be added to the system by cabling the branches together in threes (triangular).

When more than one cable is being installed on the same limb, hardware should be spaced at least 12 inches apart if practical. Hardware should never be installed in vertical alignment, directly above or below other hardware. Only one cable should be attached to each bolt or lag.

It is important to understand that most of the holding power of the bolt or lag is from the wood and callus that forms along the hardware after installation. The wood inside the branch at the time of installation may decay, thus reducing holding capacity. The new wood grows around the hardware and holds it in place.

The installation of cables in a tree represents an ongoing responsibility. Cables should be inspected annually. The hardware must be checked to see that it remains securely anchored. Loose cables should be adjusted or replaced. As the tree grows older and taller, new cables may be required higher in the tree. Trees that have been cabled may need to be pruned periodically to remove excess weight.

Lightning Protection for Trees

A bolt of lightning can destroy a tree in a fraction of a second. Trees can be blown apart, stripped of their bark, or set on fire by lightning strikes. Sometimes trees appear to have survived a strike with no more damage than a strip of bark that spirals down the trunk, but if the cambium is killed or the root system destroyed, the tree will eventually die.

It has been observed over the years that some trees are more likely to be struck than others are. Trees that stand alone in open landscapes, the tallest tree in an area, or trees standing on a hill are all candidates for lightning strikes.

Trees can be protected from lightning damage by installing a system of copper cables and a ground rod. Though sometimes expensive to install, the cost is small relative to the value of some mature trees. Golf course trees are the leading candidates for lightning protection systems for two reasons. First, some large trees are vital to the design of the

7.24. Golf course trees are good candidates for lightning protection systems because people seek shelter below them during rainstorms, and because they are often crucial to the play of a hole.

course. Trees often create doglegs or define the fairways. Second, people seeking shelter under trees during rainstorms can be killed if lightning strikes the tree. Golfers are notorious for trying to finish "one more hole" before giving in to the weather.

Besides protecting trees that are important to the play of the course, superintendents should also consider protecting trees that stand on hilltops, within 20 feet of buildings or rain shelters, or other places players

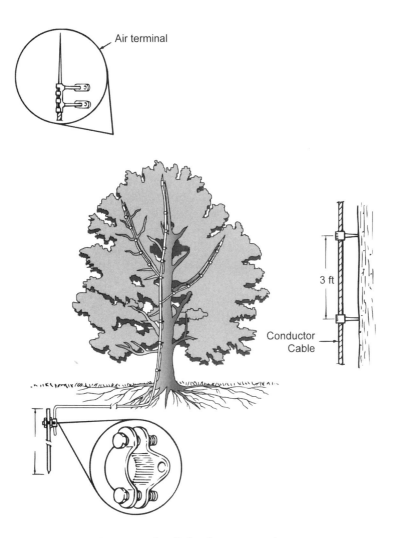

7.25. Diagram of a lightning protection system.

congregate. Protecting these trees may also minimize the liability risks associated with a strike.

A lightning protection system consists of a series of copper conductors that extend from the top of the tree, down the main branches and trunk, and out beyond the tree's dripline where they are grounded. The purpose of the system is not to prevent the tree from being struck, but rather to carry the charge safely to ground. All lightning protection hardware used in trees should be approved by the National Fire Protection Association and the Lightning Protection Institute.

The uppermost points of a system are called the air terminals. Copper or copper-bronze points are manufactured for this purpose. They are attached to the tree using barbed copper nails. They are attached near the top of the tree on the leader, if present, and on major branches within the canopy. A large tree would probably need one central air terminal and three to five secondary terminals. It is not necessary for the terminals to extend beyond the branch tips, and in fact, it is difficult to attach them to branches smaller than 2–3 inches in diameter.

The standard down conductor is a copper cable consisting of 32 strands of 17 gauge copper wire. The down conductor is connected to the primary air terminal and runs down the main branch and the trunk. The cable is attached to the tree at three-foot intervals using specialized attachments. Secondary conductor cable (14 strands, 32-gauge copper wire) is attached to the secondary air terminals and ties into the primary conductor. These are also attached to the tree at three-foot intervals. Con-

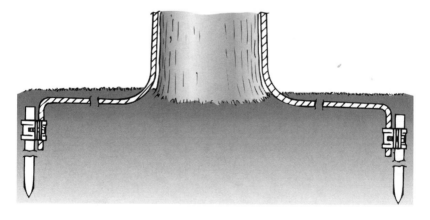

7.26. Trees that are 3 feet or larger in diameter should have two major conductors and grounds.

ductors should follow a direct path down the tree and should be attached allowing enough slack for tree sway. Trees that are greater than three feet in diameter must have two standard down conductors on opposite sides of the tree. If there are cables or wiring in the tree, it should be tied into the lightning protection system.

The conductor must be properly grounded for the system to be effective. The standard down conductor is buried about a foot below ground level, and should extend out from the tree beyond the dripline. Determining the direction the conductor extends from the tree requires some forethought. Ground rods should not be installed near underground utilities or storage tanks, nor should the conductor pass over them. Grounding cables should be attached to metal irrigation systems, but can do damage to nonmetal systems. As a rule, it is better to run the cable downhill than uphill, and toward water (ponds, creeks) is preferable.

The ground cable is attached to a copper-clad or copperweld ground rod that is 10 feet long and 1/2 inch in diameter. The ground rod should be driven into the ground to a depth of at least 10-1/2 feet. If the ground-

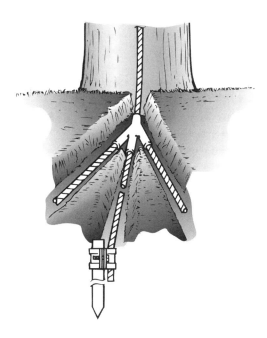

7.27. Grounds should be forked in dry, rocky soils.

ing rod reaches bedrock and cannot be driven any deeper, it should be cut off below ground level, and the remainder driven into the ground at least six feet farther out. The second ground must be attached to the first with the buried ground cable. In dry, sandy, or rocky soils, the cable should be forked to several ground rods. Grounding resistance of 50 ohms or less is considered acceptable.

If two or more trees in close proximity are protected, the grounding cables should be connected. Installing systems in the largest trees in a group can protect a cluster of smaller trees.

As with cables and braces, lightning protection systems should be inspected annually. When the tree has grown significantly past the air terminals, the conductor cables and air terminals may need to be extended. Eventually the tree may grow enough in girth to envelop the conductors. The system will still be functional as long as there is no break in the conductor or attachments. (This could, however, be a very unpleasant surprise for an arborist removing the tree some day.) Annual inspections should include a check of all splices and connections to ensure a continuous flow of current.

Summary

1. It is common for trees to be considered a low maintenance part of a golf course landscape. They do require proportionally less maintenance time than turf, flowers, and other features. Low maintenance, however, is not synonymous with neglect. Golf course trees may require pruning, fertilization, and at times, structural support.

2. Pruning is the most common tree maintenance procedure. Common reasons for pruning are to remove hazards, improve tree structure, provide clearance for buildings or people, and to increase light and air penetration to the landscape or turf below.

3. Routine pruning does not necessarily improve the health of the tree. Removal of foliage through pruning reduces photosynthesis and thus reduces growth and stored energy reserves. In order to manage trees on golf courses, however, pruning is often required. Proper pruning, with an understanding of tree biology, can maintain good tree health and structure while enhancing the aesthetic and economic values of the course.

4. Fertilizing a tree can increase growth, reduce susceptibility to certain diseases and pests, and under certain circumstances, can even help reverse declining health. If the fertilizer is not prescribed and applied wisely, however, it may not benefit the tree at all, and it may adversely affect the tree.

5. It is important to recognize when a tree needs supplemental fertilizer, what nutrients are needed, and when and how it should be applied. Fertilizer types and application techniques can greatly affect a tree's response. The advantages and disadvantages of each should be evaluated before treatments are selected.

6. Mature trees have two dimensions to survival: health and structural stability. A tree can have a healthy, vigorous crown and root system, yet be on the verge of collapse. If structural flaws are found, the installation of mechanical support (cables and/or braces) is one option for preserving trees and reducing potential hazards.

7. A bolt of lightning can destroy a tree in a fraction of a second. Golf course trees are good candidates for lightning protection systems for two reasons. First, some large trees, which have a high probability of being struck, are vital to the design of the course. Second, players seeking shelter under trees during a storm can be injured or killed if the trees are struck.

8. Trees can be protected from lightning damage by installing a system of copper cables and grounding rods. The cost of installation is small in comparison to the value of certain trees and the risk reduction.

8

Tree Hazard and Liability Issues

One of the benefits of having trees on a golf course is the mitigation of potential hazards created by golfers hitting errant shots. Trees located along the boundaries of a course can protect adjacent properties and roads from wayward golf balls. Trees between the fairways not only define the holes, but also help protect golfers from the hooks and slices hit by their fellow players.

Yet the trees themselves can present some inherent dangers. Trees can fall over or drop large limbs that can cause significant damage, personal injury, or worse, a fatality. The golf course owners and management, and the people in charge of caring for the trees, have a duty to care for the staff and players on the course. In fact, golf course superintendents, since they are presumed to have knowledge of the grounds and proper maintenance practices, may be held to a higher standard of duty than general citizens.

Caring for trees on a golf course can be viewed as a process of risk management. All trees pose a certain degree of risk. As the trees grow larger and more mature, the level of risk increases. Obviously superintendents do not want to cut down all the large, mature trees on the course. Therefore trees must be managed at a level of risk that is acceptable. Since the superintendent bears responsibility for any decisions made regarding the trees, those decisions must be based on the latest information and technology available.

Identifying Potential Hazards

The ability to predict tree failure is limited at best. You cannot always see defects, especially those inside the tree or beneath the ground, and the forces of nature are quite unpredictable. With experience, however, you can come to understand patterns of failure that will help you recognize risk factors. For example, forest edge trees are typically more prone to failure than those deeper in the woods, surrounded and protected by other trees. Look for characteristics of these trees such as bowed trunks and asymmetric crowns that may be indications of structural problems. In time you will become familiar with structural problems and defects that may lead to tree failure. The key is to continually inspect the trees under your care so that you can detect, correct, and mitigate any problems before an accident occurs.

8.1. A thorough hazard evaluation of this tree could have predicted its failure. Included bark between the codominant stems is a common defect.

General Inspection

Inspecting trees for hazards involves a great deal more than a casual stroll among them. Tree inspection must be a systematic process. It is important to understand your goals and to recognize your limits. Diagnosing tree hazards requires a fundamental knowledge of tree structure and physiology. It takes a trained eye to discern the difference between a minor flaw and a possible hazard. Understanding the level of risk management is essential in deciding what action to take when a problem is detected.

When performing an inspection, you should stick to a systematic and constant process. This will help assure that you inspect every portion of the tree and do not miss any hazardous condition. First, assess the tree as a whole. Look for dieback in the crown. Take note of any lean or branches that extend beyond the rest of the crown. Then inspect the trunk, the root crown, and the root zone. Finally, examine the canopy of the tree. By sticking to a system, you are more likely to perform a thorough inspection.

Structure

- Examine the branch angles and branch attachment of the major scaffold branches. Branches should be smaller in diameter than their parent branches. If they are equal in size or larger than the parent branch, they are much more likely to fail. Check for proper branch bark ridge formation. This is an indication of a well-attached branch.
- Watch out for codominant stems. These unions have a high probability of failure. Codominant stems often have included bark within the branch union, making them structurally unstable.
- Stems with good taper (significantly larger at the base and smaller toward the end) tend to be stronger than those without good taper. Trees that have grown in dense shade such as a forest tend to have long, straight stems with little taper. If the protection of the surrounding trees is removed, they are more likely to fail.
- Many trees have a natural lean. Trees often produce reaction wood, providing more stability to compensate for the lean. Some trees, however, may lean due to recent upheaval and could be

unsafe. Also, a lean combined with other defects may be cause for concern.

• Look for signs of vigorous growth. Some tree species are better compartmentalizers of decay than others are. Growth rate and good compartmentalization are important in the reduction of hazards. A tree that closes and compartmentalizes wounds rapidly may be less likely to become a hazard than one that does not close wounds. Although compartmentalization is important, it does not guarantee structural stability. Extensive decay, compartmentalized or not, is a risk factor.

8.2. These conks are the fruiting bodies of wood decay organisms
inside the tree. They indicate internal decay
and a potential hazard.

Potential Problems (Faults and Defects)

- Dead branches within the canopy of a tree are probably the most obvious potential hazards. The risk of damage or injury depends on the size of the dead branch and the location of the tree.
- Watch for longitudinal cracks or splits along the trunk or major branches. Cracks that start at a branch union may be especially hazardous.
- An observant arborist may be able to detect internal defects where the stem is out of round. These may be caused by the tree's formation of reaction wood or may be an indication of an internal problem.
- Externally visible defects include cankers, galls, and wounds. Each could be minor or the start of a significant problem. If such defects are discovered, further investigation is warranted.
- Some defects are man-made. Wounds, weak or damaged limbs, and decay may be the result of poor pruning or other misguided practices. Root loss, not always obvious to the untrained observer, can result from construction, trenching, grade changes, and soil compaction. Any of these may indicate a decline in health or structural stability that may require action.

8.3. The fruiting bodies at this tree's base are a sign of root crown rot.

Decay and Discoloration

Decay is perhaps the most insidious potential hazard within a tree. Decay can cause significant strength loss. There are a number of formulas to estimate the strength loss of a trunk that is decayed. While the formulas vary, most experts agree that a threshold of 30–35 percent strength loss requires some action be taken. If there are large cavity openings or other aggravating factors, the threshold drops to 20–25 percent.

The problem is that decay is not always obvious upon external inspection of the tree. The tree may appear to be solid, structurally sound and may have a thick, green canopy yet can have significant decay inside. It is important to recognize signs of internal decay.

- Watch for open wounds or cavities. They may be the starting points of decay.
- Certain insects can indicate decay. Carpenter ants are usually associated with soft, decayed wood within the tree. Wood boring insects can be another indicator. Look for frass, a sign of insect activity, accumulating at the base of the tree or in crotches.
- One of the best signs of decay is the presence of the fruiting bodies of the decay organisms. Mushrooms at the base of the tree or along major roots may indicate internal decay. Conks on the trunk or major branches are a sign of the decay fungi inside. Not all decay organism fruiting bodies are persistent, however. Some appear annually and disappear later.
- Birds and bees can also be a sign of decay. Some birds nest in tree hollows, and some feed on the insects that may be present in the wood. Honeybees often make their hives in tree cavities. Small cavities are not a problem; they serve as a habitat for birds and small animals. Large cavities can be a significant defect.

If decay is found or even suspected in a tree, the next logical step is to determine the extent. In some instances the location and extent of decay is visible. More often, a little detective work is required. If the decay is in the trunk of the tree the extent can sometimes be estimated with a rubber mallet or a very small bore drill. These simple tools, however, may be inadequate in many cases. There are a number of decay detection devices available to arborists. Most are effective and reasonably accurate when used properly, but they all are somewhat invasive.

If decay is suspected at the root flare or in the major support roots, a root collar excavation may be necessary to ascertain the extent of decay tissue. Root excavations are delicate and should only be performed by a qualified arborist. If you have any doubt when assessing decay within a tree, it is best to call in a consulting arborist who specializes in hazard assessments.

Risk Assessment and Hazard Reduction

In order to competently assess potential tree hazards, the person inspecting the trees must be trained in hazard recognition and evaluation methods. This requires not only a strong foundation in tree biology but also a fundamental comprehension of the structure-function relations of trees. Without this, potential problems may go unrecognized or inadequately evaluated.

If you are setting out to inspect trees on a golf course for potential hazards, you should proceed according to a plan, using a systematic method. First, you must have a clear understanding of your goals. The primary goal is usually to identify hazards so that they can be reduced, thereby making the course safer for all. There are limits, however, to

8.4. Root rot such as this can result in total tree failure.

the intensity of the inspection and the budget for follow-up. Whether operating in-house or hiring professionals, the individuals doing the assessment and the management must agree on the level of risk to accept. Keep in mind that the owners/managers must exercise due care in the process.

While it would be nice to perform a detailed scrutiny of each tree on the property, financial and staffing resources often prohibit thorough hazard tree inventories for even the most exclusive courses. Therefore the trees must be categorized and given priority for their level of inspection. Trees in more remote locations will be given lower priority than those along fairways, greens, or tees. Larger, more mature trees should have more thorough evaluations since they are more likely to have defects and their size can present a higher potential for damage. Give extra attention to trees in areas where people congregate.

A strategy should be developed for the assessment process. All observations, measurements, and recommendations must be documented. It is best to use a hazard assessment form for consistency. The advantage of using a standard form is that you are less likely to miss any tree or site factors that could affect the evaluation.

8.5. After this large limb failed, the decay in the branch union was obvious. The tree had never been inspected.

If you have a detailed tree inventory with the trees marked and labeled on the course master plan, you are several steps ahead at this point. Each tree surveyed and evaluated must be documented on the plan. The assessment form should identify the tree's genus and species as well as the common name. Also, record the size, location, and a general description. Note the surroundings and describe any potential "targets" that could be affected in the event of tree failure. Consider the environment of the tree including prevailing winds, microclimates, surrounding trees and other plants. Document any past history of the tree including maintenance and management practices of the tree and its surroundings.

Systematically assess the tree itself, starting with the tree as a whole, then taking a closer look at its components. In many cases, you will find it necessary to investigate further than what can be accomplished visually from the ground. It may be necessary to have a climber ascend the tree for a closer look. Some trees may require a more in-depth evaluation of the root collar. Root collar excavations, however, are quite involved, and it is best to hire a consultant with expertise in that area.

Risk assessment for structural failures includes three components: the potential of the tree to fail, an environment that may contribute to failure, and a potential target. Each factor has an impact on the hazard rating of the tree.

In evaluating the potential of the tree to fail, you must consider the defects, growth habits, branch attachments, lean, and the history of the tree. It is also important to consider the size of the potential failure. Obviously, the damage potential is greater for large limbs than small branches. The environment also plays a part in the potential for tree failure. Surrounding trees, location, prevailing winds, snow and ice loading, lightning, irrigation, and other management practices all should be taken into account.

By definition, if there is no potential target to be damaged, a tree cannot pose a hazard. Targets may include structures or people. Structures are relatively easy to assess since they are fixed. However, in determining the likelihood of a failed tree or branch striking a person, you must consider the immediate environment, traffic patterns, and use. A hazardous tree in the fairway is of much higher concern than one along a relatively obscure boundary. Yet even in remote locations there is some potential for a human target.

There are a number of ratings systems established for evaluating the hazard potential of trees. Matheny and Clark established one of the best known in *A Photographic Guide to the Evaluation of Hazard Trees in Urban Areas.* This system uses a numeric formula to quantify the risk of each tree. (The book includes a sample, standardized form for risk assessment.) The Matheny and Clark system, or other ratings systems, can be integrated into a risk management program by prioritizing the trees as to their need for attention, hazard potential, and exposure.

Mitigation Options

Part of risk assessment is evaluating the hazards and making recommendations for abatement. Some trees will pose an unacceptable risk of overall failure and will need to be removed. On a property with many mature trees, though, even the immediacy of removal may have to be prioritized. If a tree has been condemned due to structural hazards, no competent manager will want to delay its removal—the liability risk is too great. On the other hand, if dozens of trees have been slated for immediate removal, you have to proceed with those that received the greatest hazard rating first.

Sometimes there are management options short of tree removal to abate the hazards. Dead or broken branches can be removed from a tree's crown. Pruning can reduce overextended branch end weight, decreasing the likelihood of failure. If a tree has a split or codominant stems, removal of one side may be an option.

It may be advisable to install cables or braces to support weak branches, perhaps in conjunction with pruning. Such branches and any hardware installations should be checked regularly since they cannot ensure safety. If a tree is crucial to the play of a hole, a lightning protection system could be installed for additional protection against lightning damage or death.

Treatment options for cavities or hollows are limited. Past practices involved removing the decayed wood and filling the cavity with various rigid fillers. More current research indicates that filling cavities may do more harm than good. Further, decay usually develops in the interface between filler and tree. In addition, the fill material may not strengthen and support the tree as much as the new callus growth that develops around the wound. If a tree has good vitality, it may maintain structural integrity by production of new wood that forms after injury. Removal of decayed wood from a cavity has little effect. However, if healthy wood tissues are damaged in the attempt to remove the decay, the tree's ability to wall off the spread of decay may be reduced. In most cases it is better to leave the cavity alone. If the tree is not vigorous, and the cavity severe, the tree may need to be removed, or reduced significantly in size.

Another option for trees of lower risk is continued monitoring. With good management practices such as proper irrigation, fertilization, aeration, and mulching, a tree may respond favorably and improve in condition. An example would be good callus and wood production following wounding. Trees that have been pruned, or had cables and/or braces installed should also be monitored, and reassessed at some point.

Risk evaluation cannot be a one-time occurrence. All of the trees on the course should be put on an evaluation cycle that takes into account their size, age, overall condition, and the course's management practices. The frequency of evaluation depends on the budget and the level of risk that is acceptable to course management.

Legal Issues

Lawsuits and their avoidance have become a part of our every thought process. Now, more than ever, it is essential for golf course superintendents to have a basic understanding of the law, especially as it pertains to liability. It is important to know the statutes that apply in your state or province, the safety and pesticide regulations, the legal precedents for major court cases that have involved golf courses, and the potential liability that arises from poor or no management decisions. Few golf courses can afford the luxury of having an attorney on staff or even on retainer. Thus, having a basic understanding of legal issues becomes paramount.

Civil Law: Contracts and Torts

The area of law that has the greatest significance to golf course superintendents is civil law. Civil law is the set of rules which regulates relationships among people. Two of the most common areas of civil law are contracts and torts.

A contract is a voluntary and binding agreement between two parties. The courts will enforce a contract if it is consistent with public law. The goal of the courts is to protect the expectations of the parties. If the court rules that a contract has been breached, the remedy sought is to place the plaintiff in the position he would have occupied had the contract been performed. Most golf course managers enter into contracts on a regular basis. Contracts are formed with subcontractors, suppliers, and even members. Contracts do not have to be in writing to be binding, but without a written contract, the terms may be in dispute if a problem arises.

Sometimes there are damages resulting from a breach of contract. In order for the courts to award compensatory (monetary) damages, the damages must have been reasonably foreseeable at the time the contract was entered. However, the victim is expected to take all reasonable measures to minimize the damages. This is called mitigation of damages.

Many of the cases that have involved golf courses in lawsuits have arisen out of the area of law known as torts. A tort is an act other than a breach of contract that causes "injury" for which the law recognizes a right to relief. Examples of intentional torts include assault, battery, false imprisonment, slander, and libel. If a court case is brought to trial, the function of the court is to compensate the victim such that he is returned to his preinjury condition or to make him "whole." This, of

8.6. Trees along the perimeter of a golf course can serve as a buffer, providing protection to adjacent properties. When limbed up, however, the effectiveness of the buffer is decreased.

course, is not always realistic since monetary compensation cannot actually restore a person to his original status. Also, since tort cases are often tried by jury, the rulings can be inconsistent, ambiguous, and without clarification of the decision.

Liability and Negligence

When a person has suffered injury or damages, the first legal step is to place liability. Often liability is based on negligence. Negligence is the failure to exercise due care. Negligence occurs when somebody fails to perform a duty or obligation recognized by law for the protection of others against unreasonable risks. An example of negligence might be if a tree trimmer cuts branches from a tree, injuring a duffer in the rough below. In this situation the tree trimmer would be found negligent because he did not take proper precautions to protect individuals or keep them out of the vicinity. His employer will likely also be liable, and the golf course may be too, even if the tree company was contracted to do the work.

An action of negligence involves four essential elements in order for an individual to recover damages. The first requirement is the existence of a duty owed by the defendant to the plaintiff; second, the defendant fails to discharge that duty; third, "injury" results; and fourth, the injury is proximately caused by the failure to discharge the duty of care. In short, the courts must determine whether the defendant acted as a reasonably prudent person under a given set of circumstances.

Liability is based on cause. Causation in fact means that the injury can be traced back to the defendant's action (or lack of it). For proximate cause, the injury has to be reasonably foreseeable. Thus, a person cannot be held liable for unforeseen circumstances.

Acts of God

Individuals or parties are generally not held liable for injuries resulting from events that are determined to be "Acts of God." There are many legal precedents for landowners using the Act of God defense in cases where trees have fallen or broken and caused damage or personal injuries. Many times this defense has not been accepted. The reason lies in the definition of an "Act of God." The courts have defined it a number of ways, but the principle remains the same. An Act of God is an occurrence due to natural causes that could not have been prevented by ordinary skill and foresight. Thus, individuals will not escape liability for damages resulting from a fallen tree or branch if the defect that caused the failure was known of or should have been known to exist. Golf course owners/managers have a duty of care to the employees and players on the course, and must exercise due diligence in inspecting and caring for the trees on the course. If a tree fails and causes injury, the Act of God defense will not be applicable if it can be shown that the tree was structurally unsound or otherwise defective, and should have been remedied or removed. Further, the defense that the owners/managers were unaware of the condition may not hold up in court since golf courses are highly maintained landscapes and superintendents are presumed to possess a higher level of knowledge and skill in the care of turf, shrubs, trees, and other plants.

This section is not legal advice but is a survey of information stemming from many tree-related cases. Seek tree-literate legal advice to help set management risk levels and to prioritize management activities versus risks. And, of course, obtain legal advice if an accident occurs.

Summary

1. One of the benefits of having trees on golf courses is the mitigation of potential hazards created by golfers hitting errant shots. Yet the trees themselves can present some inherent dangers. Trees can fall over or drop large limbs that can cause significant damage, personal injury, or worse, a fatality. Golf course owners and management, and the people in charge of caring for the trees, have a duty of care for the staff and players on the course.
2. Caring for trees on a golf course can be viewed as a process of risk management. Since almost all trees pose some risk, and since removal of all of the trees is not a rational option, the trees must be managed at a level of risk that is acceptable.
3. The ability to predict tree failure is limited, but with knowledge and experience, it is possible to recognize patterns of failure. The trees on the golf course must be inspected on a regular basis to assess their condition and to check for potential hazards.
4. Risk assessment for structural failures includes three components: the potential of the tree to fail, an environment that may

contribute to failure, and a potential target. In evaluating the potential of the tree to fail, you must consider the defects, growth habit, branch attachments, lean, and history of the tree.

5. Tree defects include decay, root rot, split or cracked branches and trunks, and deadwood. Some of these can be treated, corrected, or removed. At times, the best alternative is to remove the tree.

6. Golf course managers and superintendents have a responsibility to maintain a course that is safe for its players and staff. Golf course superintendents, because they are presumed to have knowledge of the grounds and proper maintenance practices, may be held to a higher standard than general citizens.

7. Liability is based on cause. For proximate cause the injury has to be reasonably foreseeable. Thus if a tree fails, resulting in damage or injury, the Act of God defense will not be applicable if it can be shown that the tree was structurally unsound or otherwise defective, and should have been remedied or removed.

9

Training Your Own Crew

One option for tree maintenance is to do it "in-house" with your own course maintenance staff. There are advantages and disadvantages to this approach. One of the biggest advantages is the ability to respond to any need immediately. Contractors do not have to be called in or scheduled in advance. There is a certain luxury in not having to plan ahead or work around someone else's schedule. This luxury is, of course, not without a cost. Tree care requires expensive equipment such as chipper trucks, aerial lifts, brush chippers, chain saws, and climbing gear. Also, many courses simply do not have the budget necessary to have an arborist and tree crew on staff, or to train current staff to a level where they can safely and efficiently perform tree work. If workers are to perform tree care operations, extensive training is required.

Why Training Is Necessary

There are at least three reasons why specialized training is required for tree maintenance. First, the safety of the workers is of utmost concern. Tree work ranks among the 20 most hazardous professions and has been targeted by OSHA for closer inspection. Second, the compe-

177

tency of the workers correlates directly with the quality of the work, and therefore the appreciation or depreciation in tree values. Trees that have been well cared for may actually increase in value and will probably remain healthy and safe longer. Trees that have been the victim of improper care are likely to lose value and become hazardous. Finally, there is the obvious need to minimize liability for the course. A trained arborist can detect potential structural problems and recommend action to minimize potential hazards.

Safety

Working in and around trees can be very dangerous and safety must be the primary concern. Each year there are dozens of fatalities from falls, electrocution, tree felling, chain saw accidents, and other tree care related incidents. In addition, there are thousands of other serious accidents that result in permanent injury.

The Occupational Safety and Health Act mandates that all employers must provide work places that are free from recognized hazards. The purpose is to reduce occupational injury, illness, and death through the establishment and enforcement of safety standards and regulations, and the provision for mandatory training. Many states also have occupational safety and health laws.

The Occupational Safety and Health Administration (OSHA) has established many regulations that affect arboricultural work. For example, OSHA regulates tree work in the vicinity of electrical conductors. Most OSHA regulations are general in nature and govern multiple professions. For this reason, many professions and industries have developed individual safety standards.

ANSI Z 133.1 is a set of standards for tree care operations published by the American National Standards Institute (ANSI). It is intended to provide safety standards for workers engaged in pruning, repairing, maintaining, or removing trees or cutting brush. ANSI Standards are the recognized safety standards for tree care in the United States, and are often referred to by OSHA. They are developed and updated by a committee of tree care professionals, including representatives of the International Society of Arboriculture and the National Arborist Association.

The hazardous nature of tree work is also reflected in workers' compensation rates. Workers engaged in tree care operations must be reported under the appropriate codes. These rates are significantly higher than lawn care or landscaping rates. Workers' compensation rates for tree care vary significantly from state to state, but range from just under 20 percent to more than 60 percent. Employing tree workers can be an expensive proposition, particularly if there are safety problems.

Proper Care for Your Trees

By now you recognize that the trees on your golf course have value and perform a number of functions. Proper care in maintaining the trees is essential. There are several reasons why it pays to have trained personnel assessing these trees, maintaining them in good condition, and making recommendations for working around them.

Besides the safety of the staff working in and around the trees, the safety of anyone who passes under the trees must be considered. Tree limbs can weigh in the thousands of pounds and can do incredible damage when they fall. Because of this, workers must have extensive training before climbing trees or operating chain saws. The hazard is not limited to when trees are being worked on or removed. Improper cuts

9.1. Training to perform tree work includes safety and operational training on large equipment such as brush chippers.

or poor choices in pruning can lead to limb failure long after the equipment has been put away. Liability is usually based on negligence, and if damages or injury can be traced to substandard maintenance practices, the consequences may be devastating.

Improper tree care methods invariably lead to the decline of the tree. Poor pruning techniques can cause decline, decay, and sometimes death of the tree. Stress created by bad irrigation management or indirect chemical toxicity often leads to secondary pest and disease problems. In many cases, no tree maintenance at all may be better than poor quality, uninformed maintenance.

Almost all of the trees on a golf course have some value. Some, because of their locations, have a very high value. It is also possible for a tree to have a negative value. That is, trees that are hazardous or in very poor condition add nothing to their setting and represent a pending cost of removal. Poor pruning methods, such as topping, can reverse the value of a tree. One season a tree may be worth thousands of dollars. After topping, it is ugly, declining in health, and hazardous. A few seasons later, it is in the budget for removal, and you face the added cost of

9.2. Pruning mature trees requires knowledge of tree biology, pruning standards, and species characteristics. It also requires training in safety and climbing techniques.

replacing it. The cost of good maintenance is small compared to the costs involved with removal.

This is not intended to intimidate you if you are contemplating in-house tree work. In-house tree maintenance is still a viable option if your budget is big enough. This should, however, reinforce the fact that proper training is essential. Hiring qualified, educated personnel is a good start, but continued training is still important.

Basic Knowledge and Skills Required

If you have decided to do in-house tree maintenance, your current staff can be trained to do many of the tasks associated with tree care and removal. You would be wise, however, to hire at least two specialists, an arborist and a climber. If you are lucky, you may get both in one individual.

You need an arborist with the training and experience necessary to identify trees, diagnose problems, recognize potential hazards, and recommend treatments. These skills are not acquired in a few weeks of training. They come from a combination of basic education combined with years of experience. A good arborist will be familiar with all of the major pest and disease problems of trees in your area and will be able to make informed decisions about treatments. You must rely on an arborist's advice on issues of new construction, irrigation, fertilization, tree selection, planting, and mature tree maintenance. An arborist can help other staff members to develop skills such as pruning techniques and recognizing potential tree health problems.

No tree crew can be effective without a climber. It takes a certain kind of person to hang from a rope, operate a chain saw, and manipulate sections of trees that can weigh thousands of pounds. Tree climbing is a skill that requires a combination of knowledge, physical agility, common sense, and experience. Some tree climbers have had formal training at professional or technical schools. Most have gained hands-on training in equipment operation, safety procedures, and tree care skills. A good climber will be proficient in pruning, cabling, rigging, and recognizing tree hazards. A good climber is aware of the various characteristics of different types of trees, such as wood strength and branch angles.

Climbers must be in good physical condition to work in trees. Upper body strength is important, as the climber must pull his or her own weight up the tree. Agility, stamina, and endurance are tested constantly in a routine workday. Perhaps the most important asset of a successful tree climber is a healthy, positive state of mind. A climber who is not paying full attention can put all of the other crew members in danger.

Once you have the foundation of an arborist and a climber, the rest of the crew can be trained. The workers will need a basic knowledge of tree structure and physiology. Extensive training in safe work procedures and equipment operation is a must. Workers will have to become proficient in pruning, operating rigging lines, felling techniques, and ground work.

The following outline provides a sample curriculum for training a tree crew.

I. Understanding trees
    A. Tree identification
    B. Basic tree structure and function
    C. Defense against decay
    D. Tree health and stress
    E. Recognizing hazards

9.3. Training in proper felling techniques is essential for all workers who will be removing trees.

II. **Safety**
   A. Laws and regulations
   B. Personal protective equipment
   C. General job site safety
   D. Lifting
   E. Electrical hazards
   F. First aid and CPR
   G. Aerial rescue

III. **Pruning**
   A. Proper cut placement
   B. Pruning tools

9.4. Tree removal in confined spaces requires advanced rigging techniques. This type of work involves special knowledge, skills, and experience.

    C. Pruning techniques

    D. Impact on the tree

IV. Other tree maintenance procedures

    A. Aeration and fertilization

    B. Installing cables and braces

    C. Installing lightning protection

    D. Planting and transplanting

    E. Mulching and irrigation issues

V. Equipment use, safety, and maintenance

    A. Chipper truck

    B. Aerial lift

    C. Chipper

    D. Stump grinder

    E. Chain saws

    F. Other equipment

VI. Rigging and removal

    A. Working with the climber

    B. Operating ropes and other rigging equipment

    C. Felling, limbing, and bucking

    D. Cleanup

VII. Climbing

    A. Climbing gear

    B. Ropes and knots

    C. Rope placement

    D. Ascent and climbing techniques

    E. Tying in

    F. Working the tree

    G. Aerial rescue

How to Train

There are many resources available to help you train your crew. If your goal is in-house training, you can obtain training manuals and videos that will be invaluable. Another option is to hire professional trainers. Consulting/training companies can come in and conduct all of the required beginning training over the course of two or three weeks. This may not be practical if your location has no "off-season." A modification of this method is to hire specialists to conduct in-house training on specific subjects.

Hiring Professional Trainers

Hiring professional trainers is the most expensive training proposition. The advantage is that the experts can handle all of the planning, arrangements, and curriculum. They can provide everything from the equipment for training to the textbooks. The training courses can be customized to suit your needs and schedule. Since they are experienced in training, they have a sense of the time needed to bring your crew up to the level of knowledge and skill required for each task.

Seminars and Workshops

There are hundreds of tree-related seminars and workshops held all over the United States and Canada each year. A large percentage of these are sponsored by various chapters of the International Society of Arboriculture. Although these educational opportunities are not limited to members, it is a good idea to join your local chapter to stay abreast of the latest information and upcoming events. Many of the meetings

9.5. Private training companies can be contracted to visit your site and train your crew members.

and conferences also have a trade show where you can see the newest innovations in equipment and techniques. Trade shows are also an opportunity to make connections with distributors.

Other Training Resources

Two tree-related professional organizations, the International Society of Arboriculture and the National Arborist Association, have produced a number of books and videos designed to be used for training tree care personnel. This is a partial list of what is available as well as the addresses, phone numbers, and website addresses for both.

International Society of Arboriculture

P.O. Box GG
Savoy, IL 61874
(217) 355-9411
(217) 355-9516 FAX
(888) 472-8733 for placing orders
E-mail isa@isa-arbor.com
http://www.ag.uiuc.edu/ ~ isa/

Books

Arborist Certification Study Guide
Tree Climber's Guide
Arborist Equipment
Evaluation of Hazard Trees in Urban Areas
A Guide to Plant Health Care
Tree Pruning Guidelines
Trees and Development
Principles and Practices of Planting Trees and Shrubs
Plant Health Care for Woody Ornamentals
Arboriculture and the Law
ANSI A-300 Pruning Standards
ANSI Z-133.1 Safety Standards for Tree Care Operations

Videos

The ArborMaster Training Series

- Introduction to Climbing Techniques and Equipment
- Climbing Knots and Hitches
- Introduction to Belay: Equipment and Techniques
- Introduction to Ropes: Care, Construction and Techniques
- Introduction to Secured Footlock: Equipment and Techniques
- Introduction to Throwline: Equipment and Techniques
- Rigging Knots, Rope Slings and Eye Splices
- Innovations in Climbing Equipment
- Chainsaw Safety, Maintenance and Cutting Techniques (6 volume set)

Effects of Construction Damage to Trees in Wooded Areas
Avoidance of Construction Damage to Trees on Wooded Lots
Tree Health Management: Evaluating Trees for Hazard
Root Injury and Tree Health
Pruning Standards and Techniques for the 21st Century
Managing Trees for Public Safety (3-part series)

The National Arborist Association

P.O. Box 1094
Amherst, NH 03031-1094
(800) 733-2622
(603) 672-2613 FAX
www.natlab.com

Videos

Ropes, Knots and Climbing
Chipper Use and Safety
Chain Saw Selection and Maintenance
Chain Saw Use and Safety
Aerial Rescue
Professional Tree Care Safety (4-part series)

Electrical Hazards and Trees
Back Injury Prevention
Professional Pesticide Application

Other Training Materials

Home Study Program (3-part series)
Tailgate Safety Program
Climbers' Guide to Tree Hazards
Hazard Tree Quick Check Decals

Summary

1. Golf course managers and superintendents must decide whether to perform tree work "in-house" or to contract for it to be done by a professional tree care company. In many cases the decision is a combination of the two. Each has its advantages and disadvantages.

2. Proper training for tree work is essential for three reasons. First, the safety of the workers is of utmost concern. Second, the competency of the workers correlates directly with the quality of the work, and therefore with the appreciation or depreciation in tree values. Third, there is the need to minimize liability for the golf course.

3. All tree workers must be trained to work in compliance with OSHA regulations and the ANSI Z133 Safety Standards for Tree Care Operations. In addition, a basic knowledge of tree biology is important to understand principles of pruning, fertilization, plant health care, and various other tree maintenance procedures.

4. Training can take place in-house, utilizing some of the many training publications and videos that have been produced by the International Society of Arboriculture and the National Arborist Association. It is also a good idea to contract with training professionals to teach some of the skills involved with safe and proper tree maintenance. Workers should attend seminars and workshops designed to keep tree care workers abreast of the latest innovations and equipment for arborists.

10

Hiring a Professional Tree Care Company

Most golf course superintendents find themselves in a position of hiring a professional tree care company at some point. Some courses contract for all phases of tree work. Others may do the majority of work in-house and only contract for specific services. Either way, it pays off to make an informed decision before selecting a company and entering into a contract.

Tree Management Plan

There are a number of ways to approach tree maintenance, and the strategy chosen is usually based on the available resources—budget, personnel, and equipment. The portion of work done in-house will vary. Even if golf course staff performs all of the tree maintenance, the service of a consulting arborist is strongly recommended. The expertise and advice that an outside professional can offer is invaluable. Whether hiring a consultant or a professional tree service company, the important common denominator is the tree management plan.

The tree management plan begins with the inventory and assessment of all trees on the golf course. This is the basis of all tree care recommendations and is essential for long-range planning. Required tree work can be prioritized based on safety, effect on the course, the health and structure of the trees, and playability of the course. Some maintenance must be performed as needed (solving immediate problems), other maintenance can be scheduled for the off-season.

Services Available

When contracting with a professional tree care company, consider the services that the company offers. Some companies offer a full range of tree maintenance and monitoring services, while others do not. For example, many companies do not use chemical pest control, and some do not tackle large tree removal. Other companies are diversified into related landscape maintenance services such as mowing and snow removal. It is not necessary to choose a company with the broadest range of services if only a few services are required. It is more important to

10.1. Professional tree care companies have the equipment necessary to prune and care for trees. Trucks and other equipment used on a golf course should be cleaned and properly maintained. Gas, oil or other leaking fluids could cause necrotic patching in the turf.

select a company that will fulfill the needs of the course, while meeting the standards and expectations of the course management.

Most reputable tree service companies offer services including tree removal, pruning, plant health care, cabling and bracing. Some also install lightning protection. Many will plant and transplant trees, but the equipment owned may limit this service. Some companies are affiliated with other contractors that specialize in services such as large tree moving, large tree removals with a crane, or diversified landscape services. Some large corporations will even contract for all golf course maintenance from daily mowing to large tree care.

Consulting Expertise

The importance of obtaining the advice and recommendations of a consulting arborist cannot be overemphasized. Golf course superintendents can be responsible for more than 1000 trees, yet rarely is their primary training in arboriculture. Some golf courses actually have an arborist on staff, but this is usually the exception. The capital value of the trees justifies obtaining whatever expertise is necessary to maintain these unique assets. The money saved by acting on sound arboricultural advice will more than offset the fees paid to the consultant. A consulting arborist may be hired as an independent advisor with no tie to any professional tree care company. This has the advantage of having the assurance that the opinions given are based on an unbiased view without regard to future contracts or income. On the other hand, if the consulting arborist is part of the contractual services of the tree care company, there is a consistency and immediacy of care that can be expected from this link.

A consulting arborist can work with you and your staff in developing and maintaining a computerized tree inventory and assessment. This will aid in long-term planning and scheduling of tree maintenance as well as keeping records of what has been done and when. These records are critical for diagnosing tree problems, recommending treatments, and determining the history of trees if liability situations arise.

As the trees are evaluated for their hazard potential, health, condition, and maintenance needs, the information gathered should be recorded on the computer inventory. Whenever a tree is removed, pruned, treated for insect or disease problems, or any other care, this too should be recorded. Other important information includes routine turf maintenance in the area such as irrigation, fertilization, and herbicide use.

Any of these items may be a crucial piece of the puzzle later on if something should go wrong.

An arboricultural consultant can help you avoid liability issues by pointing out potential tree hazards and giving advice on how to abate those hazards. Our society is now more litigious than ever, and the days of claiming "Act of God" are all but gone. Golf course superintendents are trained professionals who know every inch of their courses. Courts are likely to rule that the golf course management either had constructive knowledge of, or should have had knowledge of potential hazards, should the question of negligence arise.

Any time construction or renovation is planned in the vicinity of trees on the course, a qualified arboricultural consultant should be brought in at the start. It is very difficult to treat trees that have been damaged by construction. It is far better to preserve the trees by preventing the construction damage in the first place. An arborist can work with the designer and contractors to decide which trees can and should be preserved, and then recommend measures to protect them.

One of the most common problems on mature golf courses is too much shade. Trees limit crucial sunlight and restrict air circulation, stressing turf and leading to problems with weeds, diseases, and reduced turf density. Golf course superintendents know that some trees must be pruned and others removed. An arborist can help decide which trees would be best to remove while maintaining the preferred species and specimens.

Unfortunately, convincing a greens committee or management board to allow you to remove any trees can sometimes seem impossible. Members often consider the trees on a course to be sacred. Your consulting arborist can be your ally throughout the process, helping you to build a logical case and defend your proposition. If the members of the board or committee can be helped to understand the biological situation, or the potential for injuring people or property, they will be more likely to see things your way.

Cost Analysis (In-House vs. Contractor)

There is no right or wrong answer about how much tree work to do in-house and what to contract out. Obviously the decision should be based on goals, resources available, expertise, and time. After these factors have been considered you should do a cost-benefit analysis.

As a rule of thumb, you have to be looking at a fairly large volume and long time period in order for in-house tree work to be more cost-effective than contracting. One of the biggest costs is equipment. Tree care trucks, chippers, and stump grinders are equipment that have been specially designed for tree work. They are expensive up front but can be depreciated over many years. Other equipment such as chain saws, ropes, climbing gear, and pruning tools are also required.

The largest expense is labor. Even if you utilize current crew members, they will need proper education and skills training in order to care for trees. Although they may be primarily employed as grounds maintenance personnel, you may have to pay much higher workers' compensation rates if they are doing tree work.

These additional expenses become less of a factor if they are amortized over many years. Also, if you are caring for multiple courses, it can be more cost-effective to do more in-house work. There are benefits to this approach that can outweigh the costs in some circumstances. Scheduling flexibility is a major advantage. Tree emergencies can be reacted to immediately, and there is no waiting for a contractor to return calls or schedule work.

10.2. Small portable stump grinders are relatively inexpensive, can fit into tight spaces, and do relatively little damage to surrounding areas. It may be worthwhile for golf courses to own one.

For most golf courses it will probably be more practical to contract all or most of the tree work. This eliminates the purchase of expensive equipment that may sit idle much of the time. It will also save a great deal of time and money for personnel training, insurance, and workers' compensation, and the risk potential for course employees will not be as great.

If the trees are well cared for, the costs involved with maintenance should decrease over time. In addition, the value of the trees as capital assets will increase over time. Thus, tree care on the golf course should be viewed as a long-term investment in the course.

How to Select a Tree Care Contractor

Most golf courses maintain very high standards of services and professionalism. Golf course superintendents are perhaps the most highly educated sector of the "green industry." Superintendents are trained in agronomy, horticulture, or other landscape related fields, and most have a good background in botany. With this foundation, it is reasonable that

10.3. Large stump grinding machines are pulled behind a truck. Golf course superintendents may wish to contract for this service.

golf course superintendents have high expectations of their tree care contractors. There are several criteria that should be considered when selecting a professional tree care company:

Membership in Professional Organizations

Membership in professional organizations such as the International Society of Arboriculture (ISA), the National Arborist Association (NAA), or the American Society of Consulting Arborists (ASCA) demonstrates a willingness to stay up-to-date on the latest information and techniques.

ISA Arborist Certification

Many professionally active tree care firms encourage their employees to become ISA Certified Arborists. ISA Certified Arborists are experienced professionals who have passed an extensive examination covering all aspects of tree care. In addition, in regions where it is available, many companies also employ Certified Tree Workers.

Certificate of Insurance

A current, valid certificate of insurance should be a requirement of any contractor brought onto the course. Establish a minimum coverage that is acceptable and confirm that all contractors meet that requirement. Reputable companies will usually meet such requirements anyway and will be happy to submit the certificate. It is also a good idea to request proof of Workers' Compensation coverage.

Golf Course Experience

While prior golf course experience is not mandatory for selecting a tree care company, it does indicate that the contractor will understand some of the unique situations involved with performing tree work on golf courses. Consult your colleagues at other local courses and ask for recommendations. This is one area where experience is likely to pay dividends down the road.

If the contractor has experience working on golf courses, you will not need to spend as much time explaining about protecting turf, staying off of greens and tees, avoiding irrigation systems, and maintaining a low profile while working in certain areas. Crews with golf course experience will be more prepared to modify their usual routines to avoid creating problems with the turf maintenance. For all of these reasons, it is also beneficial to build a long-term relationship with a contractor once you have found a good one.

If you can find a contractor who actually plays golf it can be a bonus. Experienced golfers understand the etiquette of the game. They can explain to their employees who may not play golf how to avoid aggra-

10.4. The International Society of Arboriculture maintains an arborist certification program that requires knowledge, experience, and continuing education.

vating the players. Little things that may seem obvious to your own crew, such as not starting a chain saw just as a golfer takes his backswing, may not even occur to nongolfing workers. In addition, fellow golfers will be able to spot where trees interfere with the play of the game, creating unfair hazards.

Make Comparisons

While it may not be best to accept the lowest bid, it is a good idea to get more than one bid. Examine the credentials of each company. Get

10.5. Removal of large trees is a service that most golf courses will contract out. The experience, equipment, insurance, and skill required is beyond the capabilities of many in-house crews.

an idea of the services offered, the standards employed, and the ability to adhere to your specifications. In addition, you must be comfortable with the contact person or arborist that you will be dealing with.

Standards of Practice

Reputable companies will only perform accepted practices. For example, practices such as topping a tree, removing an excessive amount of live wood, and using climbing spikes on trees that are not being removed are unnecessary and harmful to trees. The trees on your golf course represent a substantial asset that must be protected and preserved. Do not allow unqualified, uneducated, or poorly trained individuals to work on your trees.

Professionalism

Most golf courses maintain high standards for the appearance of their employees and equipment. There is no reason these same high standards should not be met by the professional tree care contractors. Companies that maintain their trucks in good condition, with no leaking fluids, a functional muffler, and a recent paint job demonstrate that they care about their public appearance. Workers should be appropriately dressed and should wear all of the required personal protective equipment such as hardhats when working on your trees.

Working with Your Contractor

When you hire a professional tree care contractor, you should draw up a written contract that lays out all of the specifications for the work to be done. The contract should include details of the nature of the work, the standards of practice that must be adhered to, the time frame for performance, and the compensation rate and schedule. It should cover the expectations for who will be on the job supervising the work as well as who on the golf course staff has the authority to direct the tree crew.

Both the golf course superintendent and the tree care contractor should have very clear expectations for the frequency and timing of the work. Some contracts cover only work to be done right away. Other contracts may be for work as needed over the course of a season, year, or several years. Some courses contract for regular maintenance throughout the year.

Timing can be a critical factor for doing tree work on a golf course. Most superintendents will arrange for major work such as multiple removals or major pruning to be done in the off-season. This may be the summer in hot climates, but is the winter in most temperate zones. Many times the work will be scheduled for when the ground is frozen in order to minimize damage to the course. During the regular or peak season, tree work is usually fit in between busy times such as tournaments and league play. Many courses have a closed day, often Monday, on which they try to accomplish major maintenance projects. It is important for the contractor to understand these types of parameters when the contract is established.

Another significant consideration is on-call service. A storm that causes tree damage can wreak havoc on a golf course. The downed trees and limbs must be cleaned up immediately before further damage is done to the turf and before the course can be reopened. Yet tree services are always swamped with calls following storm damage. If the golf course management has expectations of priority service by the tree contractor, it should be made clear before a contract is signed.

The prioritization of the tree work may be one of the first steps upon hiring a tree care contractor. Frequently the golf course is in great need of many facets of tree care. Too often the trees on the course have been neglected for many years. When this is the case, the importance of prioritizing the work cannot be overemphasized. The situation may be similar to medical triage, in which the trees are categorized by condition and urgency of attention needed. Once the immediate needs are assessed, the arborist and golf course superintendent must work together to schedule the "emergency" work as soon as possible.

Good communication is vital. Even if the terms of the contract are very specific, detailing what work is to be done, when it is to be done, how it is to be done, and who will be doing it, there will always be questions and misunderstandings. Working as a team and keeping the lines of communication open is essential to smooth operations. The contractor may need to be educated about the special considerations involved with golf course work. By the same token, it is the superintendent's responsibility to make all of those parameters known. General Douglas MacArthur once said, "It is not enough to give instructions that can be understood. Instructions must be given that simply cannot be misunderstood."

If you find a tree care contractor with whom you work well—one who understands the extra considerations required for working on golf courses—it may be a good idea to enter into a long-range commitment. This is actually beneficial to both parties. Long-term maintenance of trees can save money over the neglect/emergency call-in approach. Routine maintenance and a program of plant health care can keep trees healthy and structurally sound while minimizing the need for pesticide applications, severe pruning, or removals. You do not have to go through the trouble of locating a contractor each time you determine there is tree work to be done. The contractor can build the maintenance program into the regular schedule and can arrange to be available when needed. High standards of care can be sustained and the trees on the course will be healthier, less of a hazard liability, and more aesthetically pleasing.

Standards and Specifications

The contract with the tree service company should outline the standards and specifications for the tree work. This requires some knowl-

10.6. When trees fail following a storm, it can cause the course to shut down temporarily. It is a good idea to include emergency storm work as a contract provision with the tree care contractor.

edge of the arboriculture profession. If you are not up to speed in this area it is important to hire an arboriculture consultant to help you write the work specifications. Unfortunately, many tree care contractors do not operate according to industry standards. Some contractors are unaware that standards exist. Knowledge of and adherence to these standards are another screening tool for selecting a contractor.

The job specifications and standards should be based on the professional standards outlined by the arboricultural and other green industry professionals in the ANSI A-300 Standards for tree care operations and the ANSI Z-133.1 Safety Standards for tree care operations. Copies of these standards can be obtained from either the International Society of Arboriculture or the National Arborist Association. A consulting arborist should be able to provide you with sample specifications for pruning, fertilization, the installation of cables, braces, and lightning protection, pesticide application, and other facets of tree care. Always specify that all tree work performed must be in accordance with applicable safety standards and regulations.

Etiquette for Working on Golf Courses

This book makes frequent reference to the special considerations for working on golf courses. Golf courses are very highly maintained and manicured landscapes that also serve as sports facilities. This places the plants, especially the turf, under great stress. Yet the expectations for appearance and playability are unrivaled. Golf course superintendents have been fired for failure to keep greens and fairways up to the standards of the management. The turf and landscape conditions are often compared to other courses in the area that may or may not be dealing with similar problems.

The trees on the golf course can be the superintendent's worst nightmare. Shade leads to problems with grass density, weeds, pests, and diseases. Trees shed leaves, fruit, seeds, twigs, and other parts throughout the playing season. Caring for the trees is sometimes in direct conflict with turf care and the superintendent's hands are often tied when it comes to thinning or removing them.

The contractor must be aware of these inconsistencies from the start. Tree work usually involves the use of large, noisy equipment and trucks. Understandably, the superintendent will want to limit access of this equipment and keep it off the turf as much as possible. The contractor should be cognizant of the difficulty in maintaining the golf course turf and must be flexible in the work procedures. Trucks and chippers used on the course should be kept off all greens and tees and off fairways as much as possible. Equipment must be in good mechanical condition and well maintained for a professional appearance. A fuel or fluid leak can cause problems on the turf that can take weeks to repair. Exhaust from chippers or chain saws can burn the grass. Even excess sawdust can be a problem on the finely cut turfgrass.

The superintendent must specify the location of underground irrigation systems, sprinkler heads, and electrical conduits, otherwise damage could result from equipment or felled tree limbs. The contractor and superintendent must work together to protect the turf, landscape, and utilities. This will usually call for the tree workers to modify their regular methods and to take extra precautions while working.

The other major area of golf course etiquette relates to the game itself. Golf, unlike many sports, has retained many of the traditional codes of conduct, manners, and a certain unwritten decorum. Many clubs strictly enforce dress codes and rules that date back for generations. Tree care workers should be neatly dressed and in uniform, wearing the appropriate personal protective equipment to work safely and to look professional.

If contractors must work when the course is open for play, the workers must learn some of the civilities of the game of golf. The two main areas of concern will be noise and appearance. Tree equipment such as chippers, stump grinders, and chain saws are very loud. Workers should be aware of players in the area and try not to start equipment just as a player prepares to swing. Most superintendents will gladly suffer the added cost of delays in the job rather than face angry members or frustrated management. Tree workers must be made aware that their movement in and around the trees can be a distraction. They also must be cautious while working, since being hit by an errant golf shot is a very real danger.

Take Advantage of the Knowledge and Advice

Golf course superintendents sometimes make the mistake of hiring a tree care contractor to perform specific tasks, yet fail to take advantage of the arborist's knowledge and advice. Most superintendents' area of specialization is turfgrass even though they may also be responsible for the trees, shrubs, and even the flower beds. If you have taken the required steps to locate and hire a well-informed arborist, take the time to walk around the course and discuss the trees' condition and needs. The superintendent and the arborist should be a team, sharing the same goals and working toward the same objectives.

Summary

1. Most golf course superintendents find themselves in a position of choosing a professional tree care company at some point. Even if the majority of tree work is performed in-house, there are times when additional expertise or equipment is required. It is important to make an informed decision when selecting a company.
2. A consulting arborist can help you develop a tree management plan, and can help identify and prioritize the necessary tree maintenance.
3. If the trees are well cared for, the costs involved with maintenance should decrease over time. In addition, the value of the trees as capital assets will increase over time. Thus, tree care on a golf course should be viewed as a long-term investment in the course.
4. When selecting a tree care contractor, look for membership in professional organizations, ISA arborist certification, and golf course experience. Check references and confirm workers' compensation and liability insurance levels. Avoid companies that perform unaccepted practices such as topping trees.
5. Work with your arborist to decide what maintenance is required. Your arborist may be helpful in explaining tree care concerns to the board or greens committee. Take advantage of your arborist's expertise by working together as a team.

References

Periodicals

A Cut Above. *Golf Course Management.* 63(3), pp. 128,130, 1995.

Baker, B. Keeping Golf Course Trees on Par. *Arbor Age.* 18(2), p. 14, 1998.

Ball, J. Shady Characters. *Golf Course Management.* November, pp. 41–46,72, 74, 1997.

Beard, J.B. Trees or Turf. *Grounds Maintenance.* 27(10), October, pp. 12–14, 1992.

Brame, R.A. 1992. Time-Lapse Photography and Sunlight Penetration. *USGA Green Section Record.* 30(3), May/June, p. 19, 1992.

Coder, K.D. Designing Shade for Water Efficient Landscapes. *Grounds Maintenance.* 25(4), April, 1990.

Coder, K.D. Striking Back. *Golf Course Management.* 65(4), April, pp. 21–32, 1997.

Consulting Arborists: One Approach to Tree Management. *Golf Course Management.* September, pp. 42–44, 1989.

Dunster, J.A. Effective Tree Retention in Turfgrass Areas. Presentation at the 34th Annual Western Canada Turfgrass Association Conference. February, 1997.

Ermisch, M.J. Developing a Landscape Management Plan. *Golf Course Management.* May, pp. 62–70, 1992.

Fisher, R.F. and F. Adrian. Bahiagrass Impairs Slash Pine Seedling Growth. *Tree Planters' Notes,* Spring, pp. 19–21, 1981.

Frank, G.B. Fall Tree Planting. *Golf Course Management.* August, pp. 54,58, 1991.

Gadd, D. Effective Management of Tree Pruning. *Golf Course Management.* September, pp. 28–40, 1989.

Gadd, D. Treating Trees Stressed by Drought. *Golf Course Management.* June, pp. 86–90, 1989.

Gallant, A. Management by the Stars. *Golf Course Management.* 65(3), March pp. 142–143,148, 1997.

Ham, D.L. and J.R. Clark. Trees and Turf Should Complement – Not Compete. *TurfNews,* pp. 11–13, Special issue 1992.

Hawes, K. Step Back from the Forest...and See the Trees. *Golf Course Management.* 63(6), June pp. 92–104, 1995.

Hawes, K. Summer Tree Care and Maintenance. *Golf Course Management.* June, p. 104.

Horvath, T. and C. Schreiner. A Growing Concern: A Golf Course Architect's View of Tree Placement on the Course. *Golf Course Management.* January, pp. 237–247, 1998.

Kocher, B.G. Working with Sun and Shade. *Golf Course Management.* September, pp. 40–44, 1993.

Ling, J. Playing the Game. *Arbor Age.* 15(2), February, pp. 10–12, 1995.

Mathewes, J. Between the Fairways. *Golf Course Management.* 64(11), November, pp. 32, 36–37, 1996.

Miller, R.H. A Tool for Managing Golf Course Trees: Plant Health Care. *Golf Course Management.* 61(9), September, pp. 32–36, 1993.

Moore, J.F. Fire in the Hole. *USGA Green Section Record.* 30(3), May/June, pp. 25–26, 1992.

Nesbitt, S. Beauty and the Beast. *Golf Course Management.* September, pp. 36–44, 1994.

Oatis, D.A. Using New Technology to Solve an Old Problem: Trees. *USGA Green Section Record.* 35(3), May/June, pp. 20–21, 1997.

Reid, M.E. Effects of Shade on the Growth of Velvet Bent and Metropolitan Bent. *USGA Bulletin.* 13(5), October, 1993.

Seeking Tree Maintenance Contract with Golf Courses. *Arbor Age.* 13(1), January, pp. 10–12, 1993.

Skorulski, J. Developing a Tree Care Program. *USGA Green Section Record.* 34(2), Mar/Apr, pp. 1–7, 1996.

Smith, R.C. Understanding the Process of Mature Tree Decline. *Golf Course Management.* 61(9), September, pp. 45–50, 1993.

Snow, J.T. A Guide to Using Trees on the Golf Course. *USGA Green Section Record.* 18(4), July/Aug, pp. 1–5, 1980.

Snow, J.T. The Fall Harvest. *USGA Green Section Record.* 28(2), Mar/Apr, pp. 18–19, 1990.

Snow, J.T. Trees, Trees Everywhere. *USGA Green Section Record.* 22(1), Jan/Feb, pp. 1–5, 1984.

Tree Preservation Programs a Must. *Turf and Recreation.* 6(7), Nov/Dec, pp. 36, 45, 1993.

Trees Make Golf Course Stand Out from Crowd. *The Davey Bulletin.* July–October, 1995.

Vermeulen, P. Ten Timely Tips to Avoid Tree Troubles. *USGA Green Section Record.* 28(5), Sept/Oct, pp. 15–17, 1990.

Watschke, G.A. The Monsters of Manchester. *USGA Green Section Record.* 24(5), Sept/Oct, pp. 1–5, 1986.

Watson, G.W. Competition Between Trees and Turf. *Grounds Maintenance,* October, p. 30, 1989.

Watson, G.W. Organic Mulch and Grass Competition Influence Tree Root Development. *Journal of Arboriculture.* 14(8), August, 1988.

Watson, G.W. Tree Growth Revisited. *Golf Course Management.* June, pp. 8–24, 1990.

Watson, G. and G. Himelick. *Golf Course Management.* July, pp. 149–155, 1997.

White, C.B. Shady Characters. *USGA Green Section Record.* July/Aug, pp. 1–3, 1985.

Zimmerman, M. How to Select an Arborist. *Golf Course Management.* October, pp. 68–72, 1989.

Books

ANSI A-300. American National Standard for Tree Care Operation—Tree, Shrub and Other Woody Plant Maintenance—Standard Practices, American National Standards Institute, Inc., New York, NY.

ANSI Z-133.1. Safety Requirements for Pruning, Trimming, Repairing, Maintaining and Removing Trees and for Cutting Brush, American National Standards Institute, Inc. 1994. New York, NY, 1994.

Dirr, M.A. *Manual of Woody Landscape Plants,* 4th ed. Stipes Publishing, Champaign, IL, 1990.

Gilman, E.F. *An Illustrated Guide to Pruning.* Delmar Publishers, Albany, NY, p. 178, 1997.

Harris, R.W. *Arboriculture: Integrated Management of Landscape Trees, Shrubs and Vines,* 2nd ed. Prentice Hall, Englewood Cliffs, NJ, 1992.

Lilly et al. *Arborists' Certification Study Guide.* International Society of Arboriculture, Savoy, IL, 1993.

Lilly, S. *Tree Climbers' Guide.* International Society of Arboriculture, Savoy, IL, 1994.

Lilly, S. *Tree Worker's Manual,* 2nd ed. Ohio Agricultural Curriculum Materials Service. Columbus, OH, 1994.

Matheny, N. and J.R. Clark. *Trees and Development: A Technical Guide to Preservation of Trees During Land Development.* International Society of Arboriculture, Champaign, IL, 1998.

Matheny, N. and J.R. Clark. *A Photographic Guide to the Preservation of Hazard Trees in Urban Areas,* 2nd ed. International Society of Arboriculture, Savoy, IL, 1994.

Merullo, V. and M.J. Valentine. *Arboriculture and the Law.* International Society of Arboriculture, Savoy, IL, 1992.

Plant Health Care for Woody Ornamentals. Illinois Cooperative Extension Service and the International Society of Arboriculture, Savoy, IL, 1997.

Shigo, A.L. *A New Tree Biology Dictionary,* 2nd ed. Shigo and Trees, Durham, NH, 1989.

Shigo, A.L. *A New Tree Biology Dictionary.* Shigo and Trees, Durham, NH, 1986.

Tree Pruning Guidelines. International Society of Arboriculture. Savoy, IL, 1995.

Watson, G. and E.B. Himelick. *Principles and Practice of Planting Trees and Shrubs.* International Society of Arboriculture, Savoy, IL, 1997.

Index